The Gentle Desert

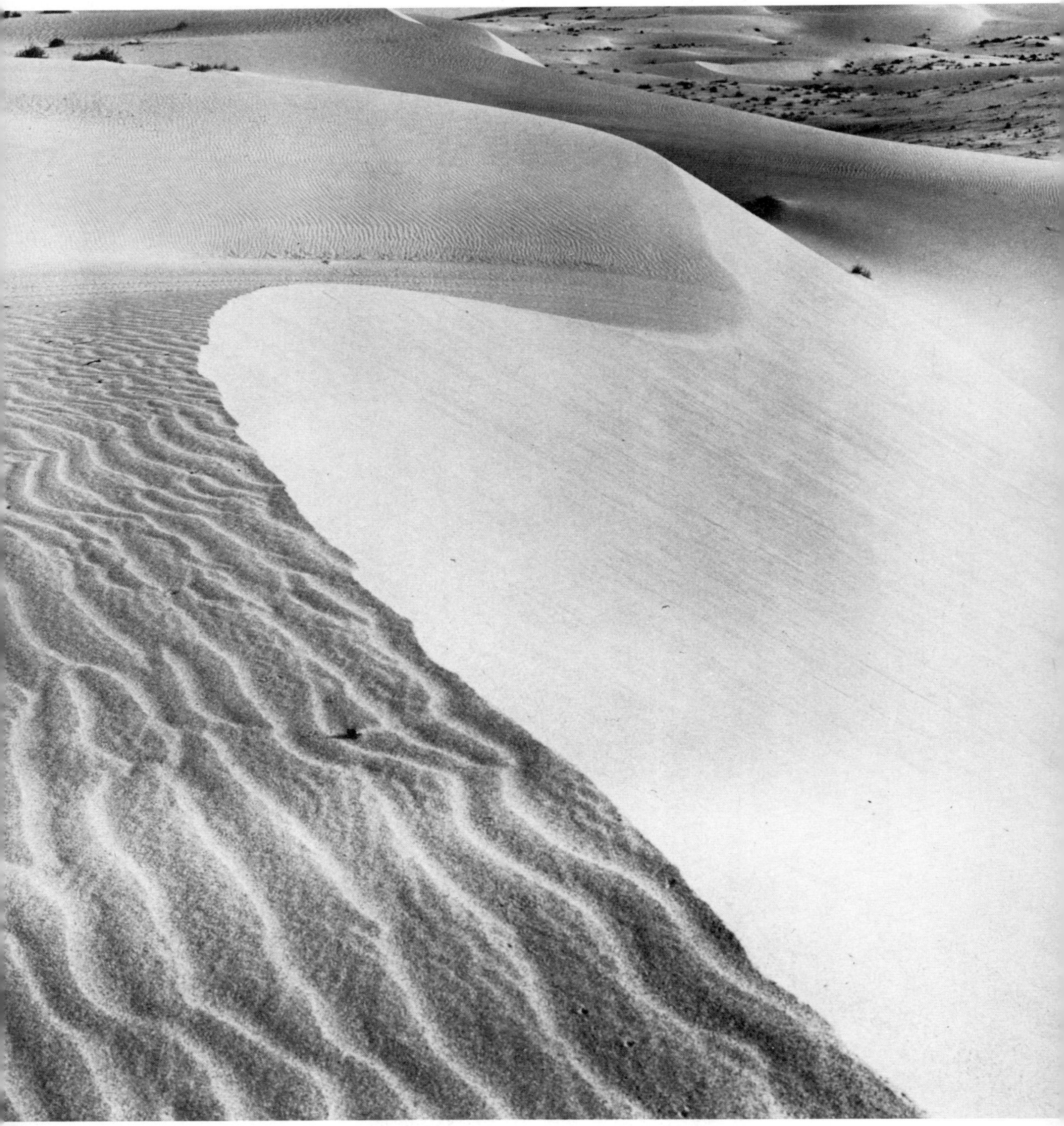

The Gentle Desert
Exploring an Ecosystem

By Laurence Pringle

MACMILLAN PUBLISHING CO., INC. New York
COLLIER MACMILLAN PUBLISHERS London

The author wishes to thank David F. Costello, author of *The Desert World,* for reading and suggesting changes in the manuscript of this book.

Photograph on page 47 reproduced courtesy of
U.S. Bureau of Land Management
Maps on pages 5 and 7 by Rafael Palacios

Copyright © 1977 Laurence Pringle
Copyright © 1977 Macmillan Publishing Co., Inc.
All rights reserved. No part of this book may be reproduced or transmitted in any form or by any means, electronic or mechanical, including photocopying, recording or by any information storage and retrieval system, without permission in writing from the Publisher.
Macmillan Publishing Co., Inc.
866 Third Avenue, New York, N.Y. 10022
Collier Macmillan Canada, Ltd.
Printed in the United States of America
10 9 8 7 6 5 4 3 2 1

LIBRARY OF CONGRESS CATALOGING IN PUBLICATION DATA

Pringle, Laurence P
   The gentle desert.

     Bibliography: p.
     Includes index.
     SUMMARY: Describes the plant and animal life of the North American desert.
     1. Desert ecology—North America—Juvenile literature. [1. Desert ecology. 2. Ecology] I. Title.
QH102.P67   574.5′265   77–5875
ISBN 0-02-775380-8

The desert is a vast world, an oceanic world, as deep in its way and complex and various as the sea.

—Edward Abbey
Desert Solitaire

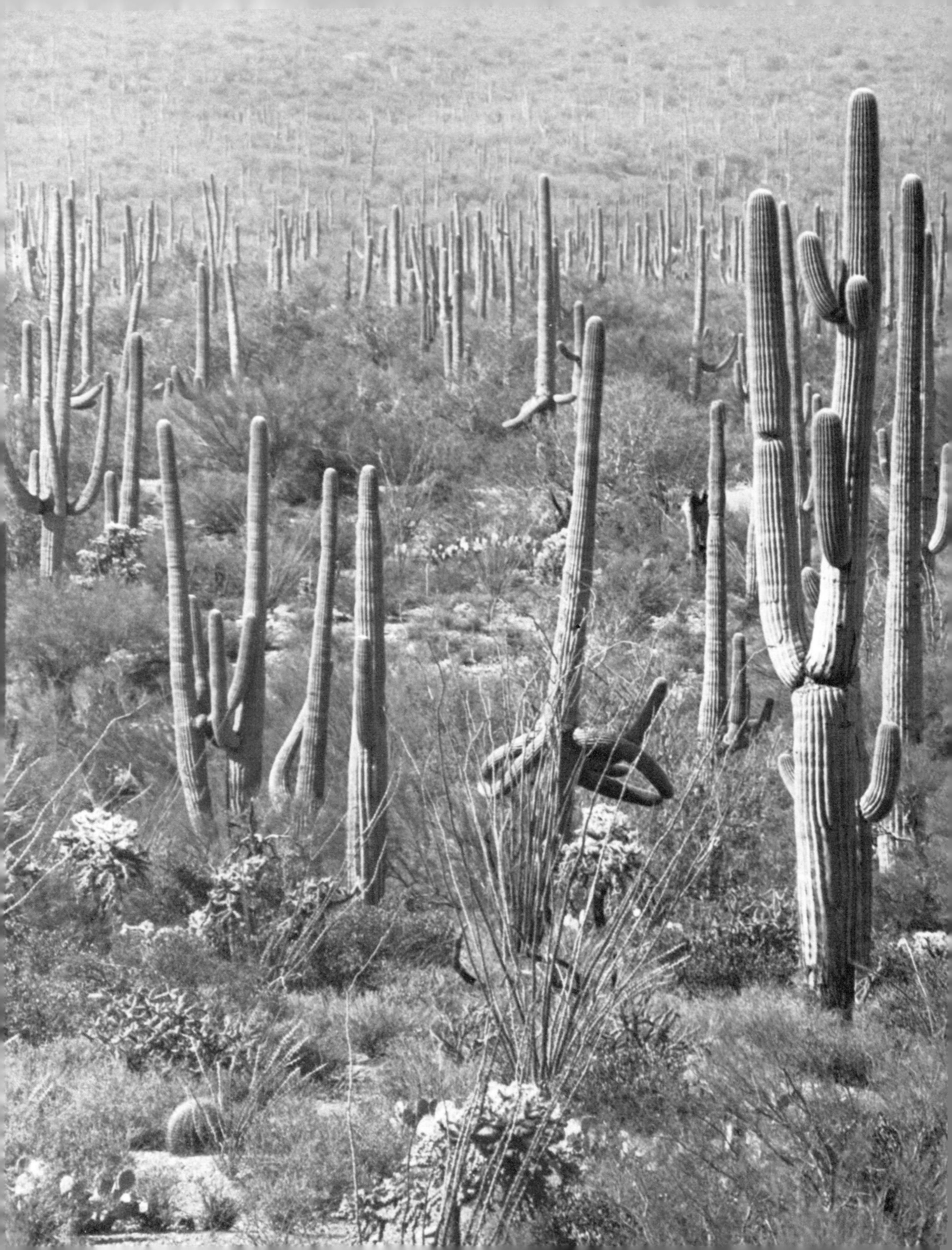

About This Book

A desert is a kind of ecosystem—a place in nature with all of its living and nonliving parts. Ecosystems are all around us. Some are big, some are little. The planet earth is one ecosystem, a rotting log is another. Forests, ponds, and backyards are also ecosystems.

This book introduces desert ecosystems. Deserts cover nearly a fifth of the land in the United States (excluding Alaska and Hawaii), yet most people know little about them. For many years people thought that deserts were useless wastelands. Now more and more people are discovering that desert ecosystems are rich with plant and animal life.

Deserts are lands of contrasts—hot days and cold nights, floods and droughts, barren dunes and cactus forests. Deserts are the home of seed-eating rodents that never drink water, fringe-toed lizards that dive into dunes, and little owls that gobble down scorpions. Deserts are also "living laboratories" where scientists study some of the wildest places left on earth.

There are many different kinds of desert around the world, but they are all alike in some ways. This book is about the North American Desert. It is the most varied desert on earth, and a fascinating ecosystem to explore.

2

Summer is vacation time for most families. It is also the season when many people see a desert for the first time. They rush along highways in the southwestern United States, hurrying from one motel or campground to the next. It is usually very hot. Beyond the car windows are endless miles of bare earth and scrubby-looking plants. It is easy to conclude that the desert is a harsh, unfriendly, and almost lifeless place.

Or so it seems at 60 miles an hour. You will get a very different impression if you walk through the desert, especially at dawn or dusk. Birds are singing. Lizards dash after insects. A jack rabbit bounds across your path. The desert is a lively place.

Still, it may seem strange and even a bit frightening to people who are used to green forests or pastures or lawns. Deserts are different. Even the light and the air seem to have a special quality.

Actually, they *are* different, because the air is usually quite dry. This lack of water vapor also makes the air clear. Often it is so clear that a distant mountain seems to be just a few miles away.

This lack of moisture in desert air can have several causes. One is the wind patterns of the earth. The winds around the middle of the earth are usually dry, having lost much of their water vapor to cooler lands farther north or south. As a result, most of the earth's deserts occur around the middle of the world.

Some deserts occur to the east of mountain ranges. The mountains act as barriers to moist air that usually comes from the west. Most of the water vapor in the air falls as rain or snow on the western side of the mountains. This leaves very little for the other side. Along western coasts, cool ocean currents have the same effect. They cause a lot of water vapor to fall into the sea as rain. One way or another, land is shut off from moisture-filled air and the result is a desert.

Why Deserts Form in Western North America

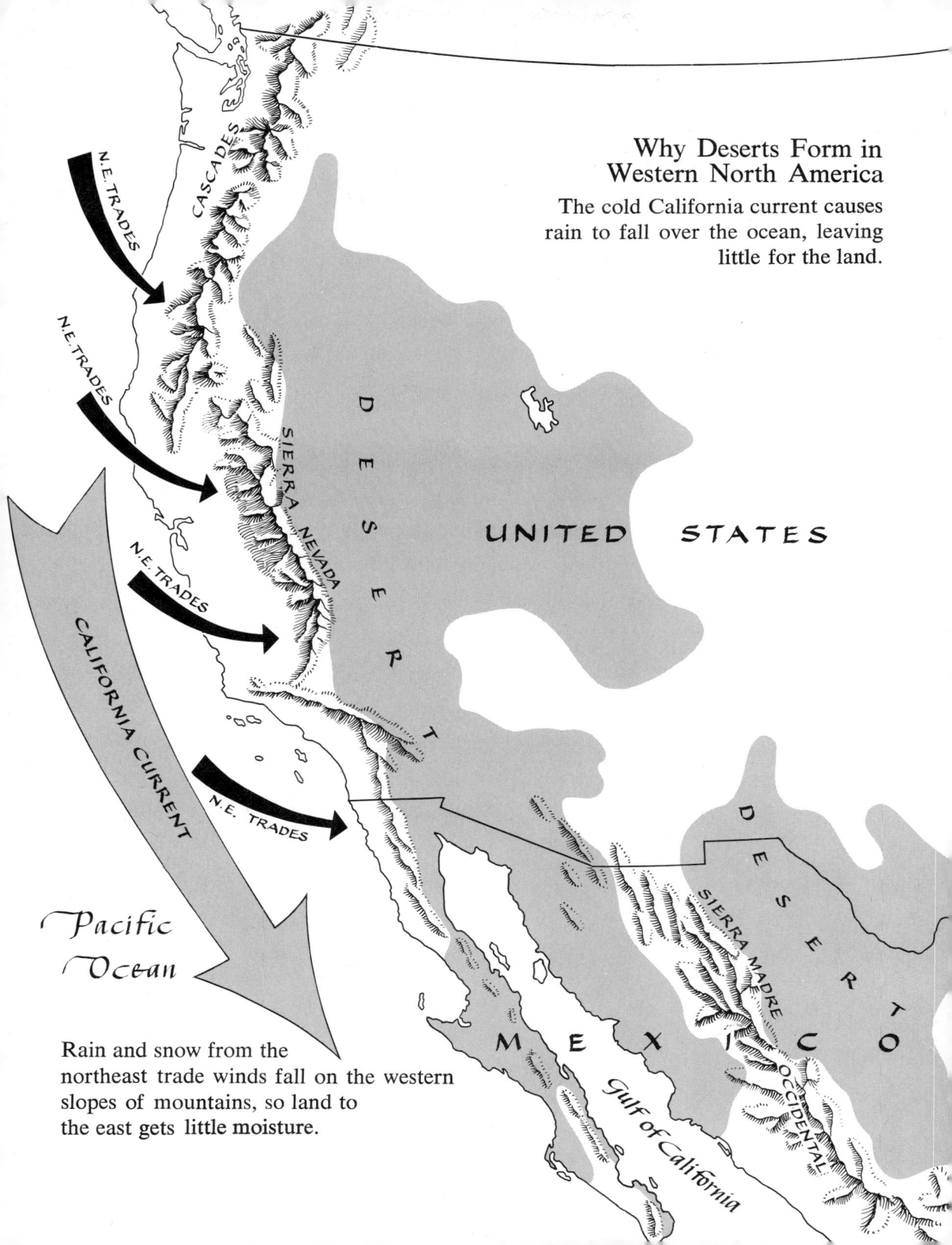

The cold California current causes rain to fall over the ocean, leaving little for the land.

Rain and snow from the northeast trade winds fall on the western slopes of mountains, so land to the east gets little moisture.

Only ten inches or less of rain (or snow) falls on a desert during a year. Moreover, this precipitation usually comes in one brief season—followed by nine dry months. Besides deserts, two other regions of the world have very little precipitation—the Arctic and Antarctic. These regions are sometimes called polar deserts. They are quite cold year round. Deserts, as we commonly think of them, are usually hot, at least for part of each year.

Deserts cover 14 percent of all the land on earth. In North America they occupy nearly 500,000 square miles in the United States and Mexico. The North American Desert reaches its northern limit in Oregon and Washington, close to the Canadian border. The southern limit is in central Mexico. You can find desert environments on mountains, 5,000 feet above sea level, in such places as Big Bend National Park, Texas, and also 280 feet below sea level, in Death Valley, California.

Although the entire North American Desert is quite dry, its climate varies a lot from place to place. The largest and northernmost part, called the Great Basin, is a cool desert. About half of its precipitation falls as snow. The time of precipitation also varies in deserts. Winter is the rainy (or snowy) season in the northern and western parts, while the southeastern regions get summer rains. The area near Tucson, in eastern Arizona, gets both summer and winter rains.

Major North American Deserts

As the climate varies, so does the desert life. In fact, some regions are so different from one another that the North American Desert is usually divided into four parts—the Great Basin, Mohave, Sonoran, and Chihuahuan deserts. They are alike in some ways, but each has distinctive plants and animals. Sagebrush is the most common shrub in the Great Basin. Farther south, the Sonoran Desert is cactus country. More than sixty species of cacti live there, including the tall saguaro (pronounced sah-WAR-oh).

Each of the four parts of the North American Desert is a huge ecosystem that covers many square miles. Each has a variety of land forms, climates, soils, and life. Each of these huge ecosystems is made up of many smaller ones. Let's get acquainted with some of these smaller ecosystems—the kind you may see on a hike through a desert.

One ecosystem is the dry wash, or arroyo—a stream bed where water flows only after a heavy rain, often just once a year. At this time an arroyo can be dangerous, for a flood can flash down a channel that is dry and safe most of the year. The channel of an arroyo is more barren than the surrounding desert. Flood waters prevent shrubs from getting a roothold there. But shrubs and trees are abundant along the edges of an arroyo, where water is close to the surface. Palo verde, smoke tree, ironwood, and mesquite (pronounced mes-KEET) are common arroyo plants.

A dry wash (arroyo) at Joshua Tree National Monument, California.

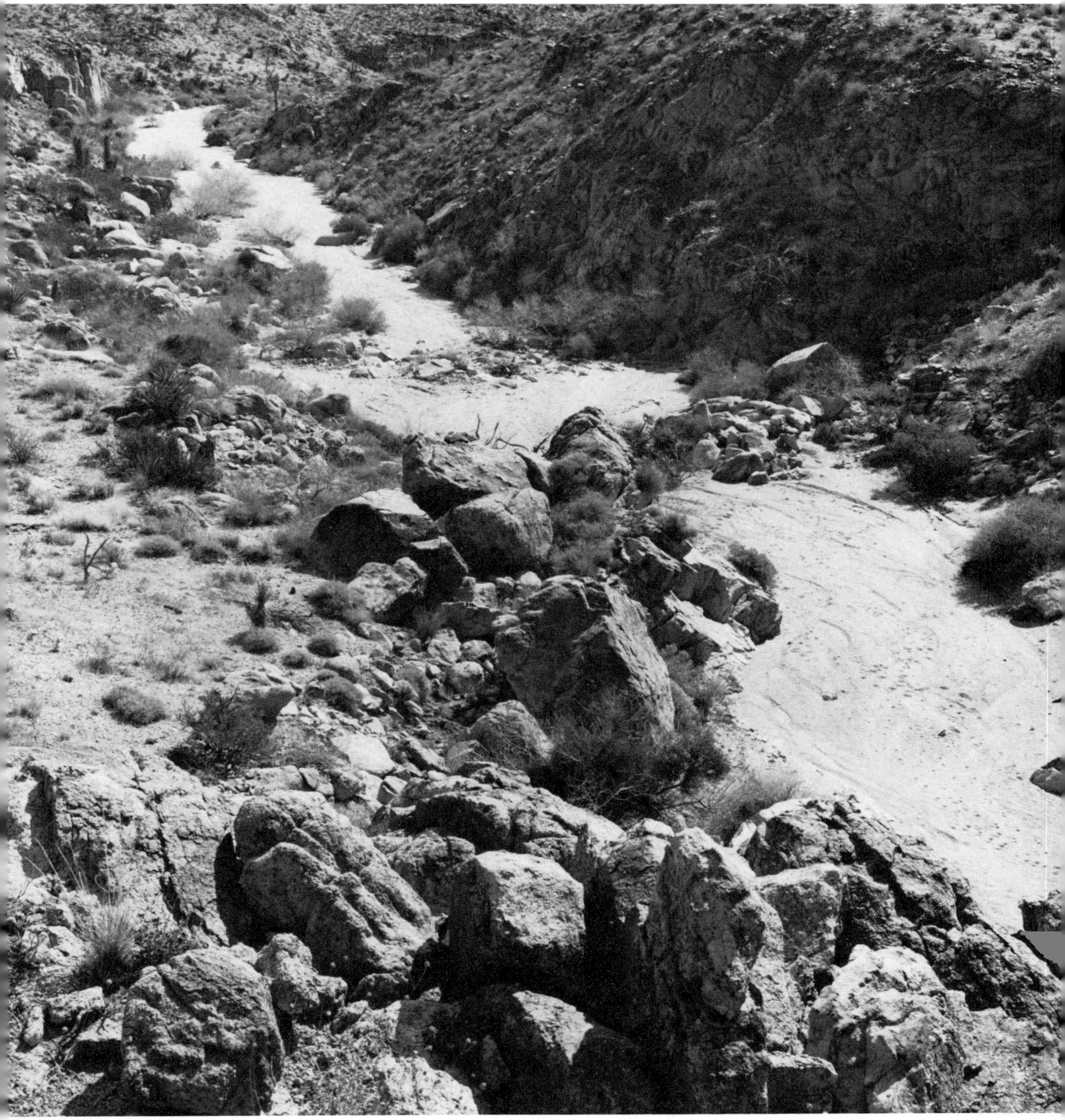

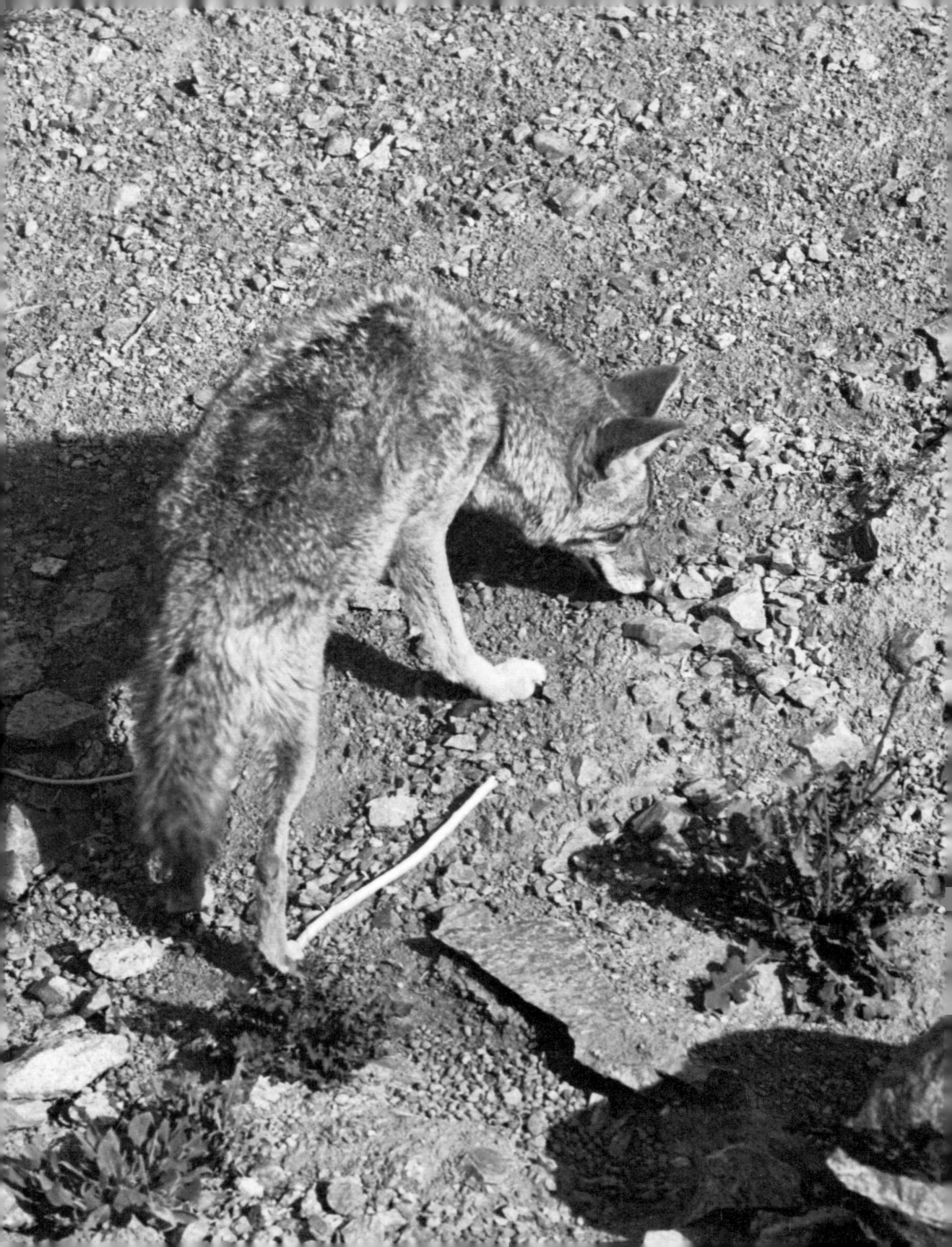

The special conditions of arroyos—especially the floods—affect the reproduction of common arroyo plants. The seeds of most arroyo trees and shrubs have tough outer coats. Left untouched, these seeds do not sprout because moisture cannot get through to the embryo plant. During an arroyo flood, however, the seeds are tumbled and battered on the rough surface of the channel. Their coats are scratched and scarred. Then water can enter, and the seeds sprout. They quickly send roots deep into the soil. A seedling only two inches high may have a root two feet long.

The large plants of the arroyo ecosystem attract many animals, including insects that get nectar from their flowers, birds that nest in their branches, and rodents that eat their seeds and leaves. Water may be only a few feet underground in an arroyo, and in the driest seasons coyotes sometimes dig down for a drink. These "coyote wells," as they are called, then attract doves, quail, mice, and other animals. Arroyos are also used as easy-to-travel paths by coyotes, deer, and jack rabbits (which are actually hares, not rabbits).

You may see coyotes (left) or their tracks in dry washes, and quail (right) among the arroyo bushes.

Sand dunes are another special desert ecosystem. For many people the word *desert* brings to mind pictures of huge dunes, stretching as far as the eye can see. Such dunes exist in parts of the world's largest desert, the Sahara of northern Africa. But you can travel a hundred miles without seeing a single dune in the deserts of North America. This continent's largest dune areas are west of Yuma, Arizona, and at the White Sands National Monument in New Mexico.

White Sands Monument is a 275-square-mile park of white gypsum sand dunes. The sand is soft to the touch, and the dunes are beautiful. At first they seem lifeless; then you notice a few yucca plants. Their roots reach deep down for water. Wherever yuccas grow, small moths also live. At night they visit yucca flowers, spreading pollen from one to another, accidentally helping the plant produce seeds. Female moths also lay eggs within yucca flowers. The larvae which hatch from the eggs are able to develop and grow because they are in the midst of yucca seeds. They eat some, but others are left and the plant can reproduce itself.

Dunes may seem lifeless in midday, but tracks in the sand reveal the presence of foxes, rodents, lizards, and insects.

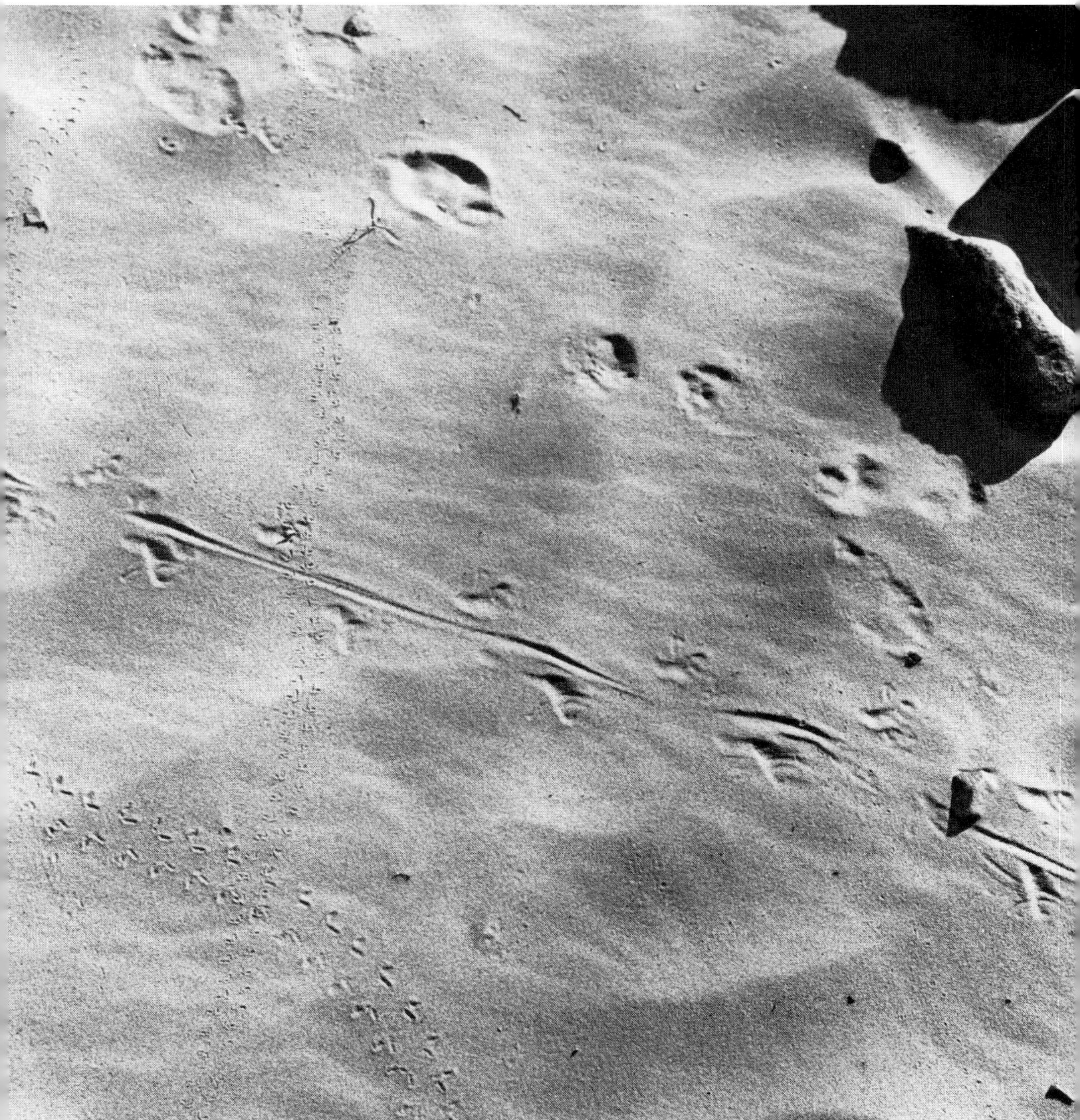

Dunes have a surprising variety and abundance of animal life. Animals cannot exist without plants, however, and the dunes with the most plants growing on them, or nearby, have the richest variety of animals. Several kinds of lizards, snakes, and rodents spend most of their lives on dunes and other sandy areas of the desert. There is even a kind of cockroach that lives on dunes. We usually picture these insects in damp places—near kitchen sinks, for example. But desert cockroaches spend most of their lives beneath the dune surface, where they find water vapor and food.

One of the most unusual dune animals is the fringe-toed lizard. Fringes of long, pointed scales on its hind feet make them broader than usual—somewhat like snowshoes. These feet enable the lizard to run rapidly over sand. It runs after food such as insects and smaller lizards. It runs away from animals that want it for food.

This lizard can even disappear into loose sand. It dives in, burrowing with its wedge-shaped head, kicking with its hind feet. The shape and smooth surface of its body help it to slip easily into loose sand. The entire disappearing act takes only a second or less. Fringe-toed lizards can "swim" a few feet in loose sand. This makes it more difficult for predators (meat-eating animals) to find them. It also enables them to reach sand that is much cooler than at the surface.

Long scales on the feet of the fringe-toed lizard (above and right) enable it to move quickly on loose sand.

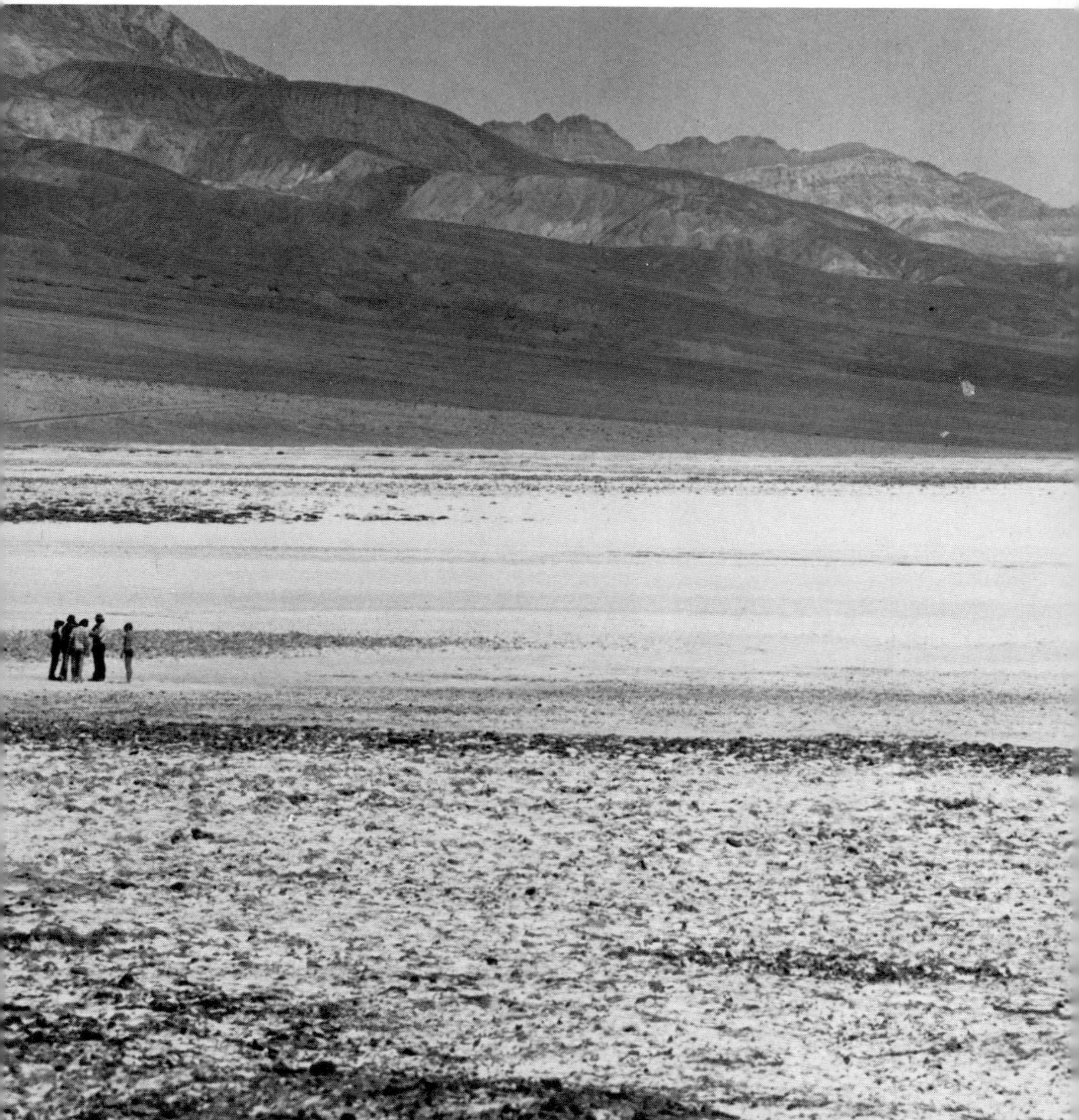

Besides dunes and arroyos, desert ecosystems include dry lakes, called playas. Like dry washes, they occasionally hold water after a rain but are usually dry. The rain water evaporates quickly. Salts in the water gradually form a crust on the surface. Only a few kinds of plants can grow in the playa ecosystem. They have roots that grow deep underground where the soil is less salty than at the surface. One of these plants is called salt bush; another is pickleweed.

Some parts of the desert are mostly rock, upon which only simple plants called lichens grow. Other plants take root in crevices and wherever a little sand or soil collects. A lush garden of plants may grow in a rocky canyon. Rain water tends to flow into a canyon, and the canyon walls provide shade.

Few plants live on this Death Valley playa (left), but a surprising variety of plants grows in the rocky and sandy environment of the desert.

This shade helps create a climate that is wetter and cooler than the surrounding land. Small differences in climate can have great effects on plant and animal life. In the Sonoran Desert, for example, tall saguaro cacti are most abundant on slopes that face south and southwest. The slopes that face north are cooler and a bit more moist. Saguaros grow very well there, but every once in a while they are damaged or killed by freezing temperatures.

Small differences in climate also affect the survival and growth of young saguaros. The saguaro seeds that fall onto bare desert soil seldom survive. Those that land in a rocky crevice or in the shade of a bush are more likely to begin to grow. The microclimate ("little climate") under such a "nurse" bush is cooler than in the open, a few feet away. When frost strikes the desert, the microclimate under a bush is usually a bit warmer than in the open.

A saguaro cactus forest is probably the most complex desert ecosystem in the world. In just a small area among saguaros you can find an amazing number of plants and animals to observe and study. The cacti themselves are the tallest objects in the desert—fifty feet high. They are favorite perching places for hawks, owls, and vultures.

Saguaros in their "nursery," under a palo verde tree. A two-foot-tall saguaro is about 25 years old.

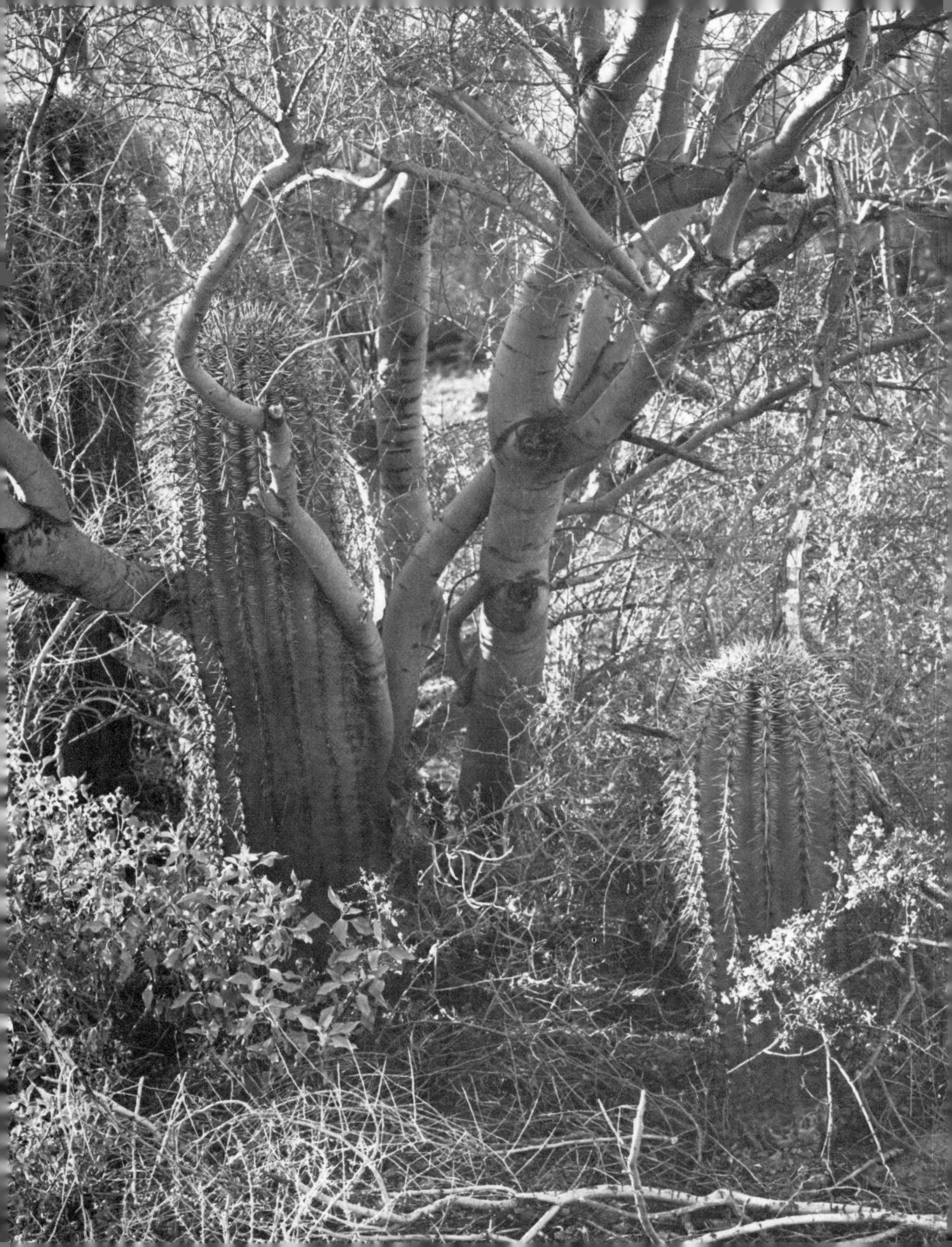

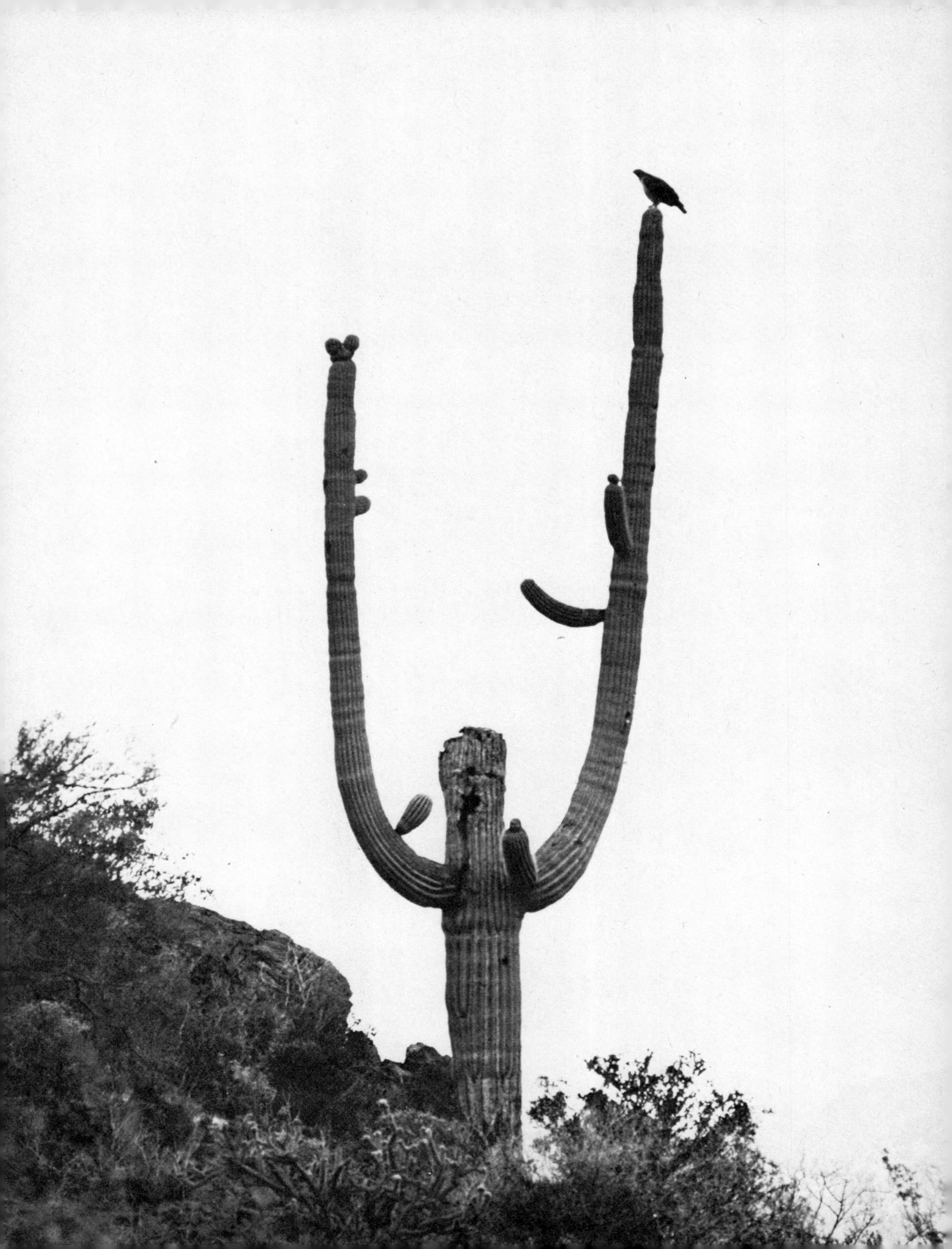

Each spring woodpeckers chip holes in saguaro trunks and raise their young inside these cavities. Since woodpeckers make a new nest hole each year, their former homes are available to new tenants. For many animals, these holes are as precious as a water hole. The microclimate inside is often more than ten degrees cooler than outside. The nest holes are also quite safe from predators because they cannot climb up the steep, spiny saguaro trunks.

At least sixteen species of birds nest in saguaro holes. One is the elf owl, the smallest in North America. It eats moths, other insects, and scorpions. Birds compete for saguaro holes with each other, and also with small mammals such as bats, pack rats, and cactus mice. Lizards, insects, and spiders also seek refuge inside saguaros. As far as we know, one species of mosquito exists solely because of these cavities. Its young develop in the rain water that collects inside some saguaro holes.

Large hawks (left) perch atop saguaros, and small owls nest inside saguaro trunks.

Saguaros bloom from April through June. Bees and moths visit the flowers for food, and accidentally carry pollen from flower to flower. The blossoms are also pollinated by doves and other birds by day and by long-nosed bats at night. The bats and birds sip nectar from the flowers. There are many flowers, and each one that is pollinated produces about two thousand tiny seeds, packed inside a bright red fruit. Only a few of these seeds survive to sprout and begin growth. Most are eaten by mice, ants, and birds.

The saguaro is a remarkable plant. Many animals depend on its nectar, its seeds, and the nest holes made in it by woodpeckers. But this tells only part of the story of the saguaro's importance in its ecosystem.

We can discover more of the story by tracing part of a food chain back to its beginnings. Suppose you see a vulture eating a dead coyote. As the vulture digests its meal, it gets some energy, minerals, and nutrients from the coyote's flesh. Like all animals, the coyote's body was made from food it ate. The coyote may have gobbled down some saguaro fruits. It almost certainly ate many jack rabbits, and these hares sometimes eat young saguaro plants. The coyote also ate pack rats and mice. These rodents commonly eat saguaro seeds. One way or another it is likely that every animal in a saguaro ecosystem—from ants on the ground to vultures in the sky—is connected to the saguaro cactus.

Turkey vultures are important scavengers of the North American Desert.

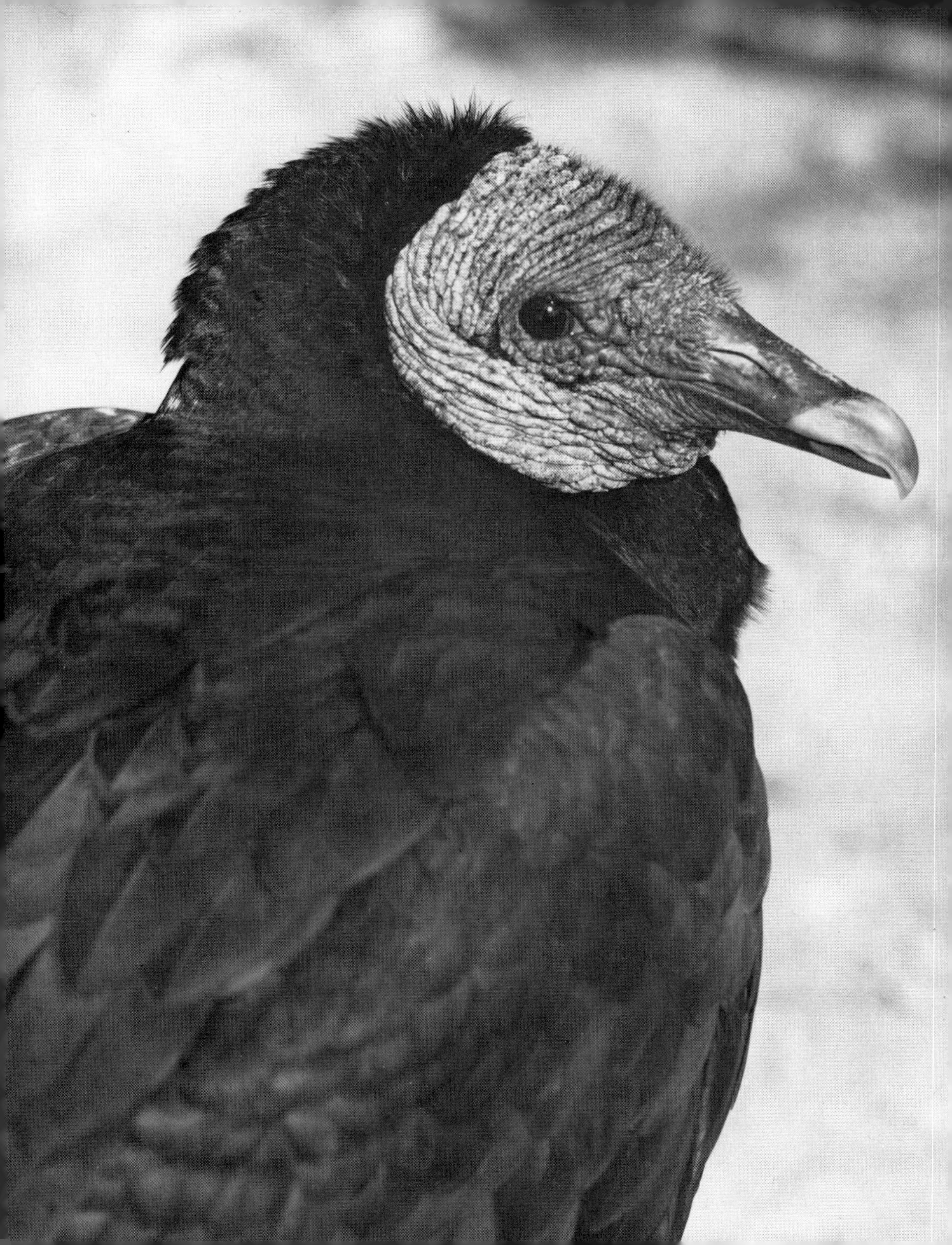

In the Great Basin Desert, animals depend heavily on the sagebrush plants that are so common there. In the Mohave Desert, another plant is a major source of food and shelter. It is the Joshua tree, a thirty-foot-tall member of the yucca family. In parts of the Mohave Desert, Joshua trees form forests as the saguaros do in the Sonoran Desert.

One lizard lives almost exclusively on Joshua trees. The yucca night lizard, about three inches long, is usually found among dead leaves on living trees, or under dead limbs that have fallen to the ground. The lizard's food—insects and spiders—also lives in the same hideouts, so it seldom leaves its Joshua tree environment.

Two biologists counted the animals they found among Joshua trees in southern California. They listed twenty-eight species of mammals, including bobcats, gophers, kangaroo rats, badgers, and several kinds of mice. Besides the yucca night lizard there were nearly thirty other kinds of lizards and snakes. Fifty different kinds of birds lived in the ecosystem, and half of them built nests in Joshua trees.

A yucca night lizard on a dead, fallen Joshua tree

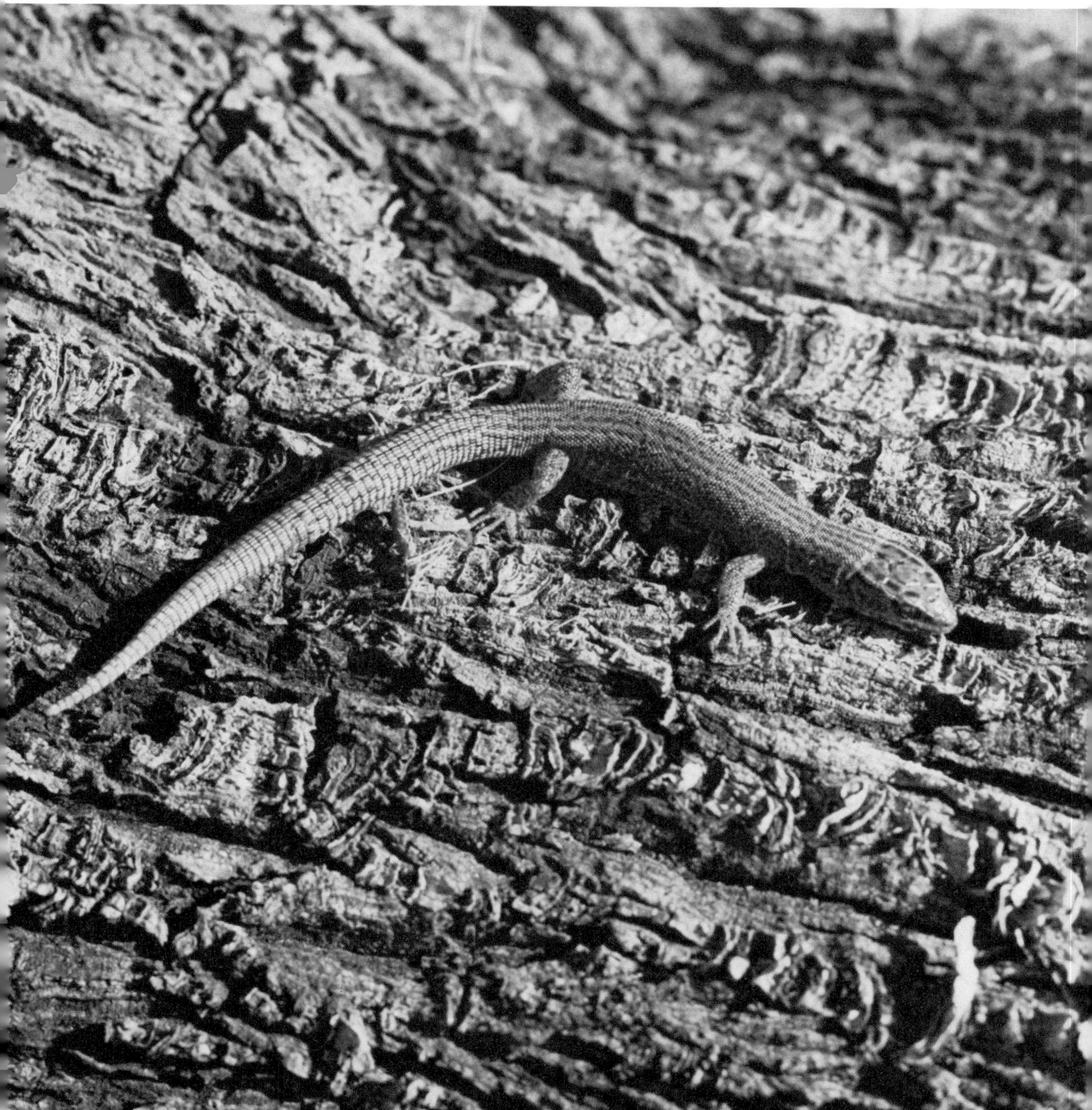

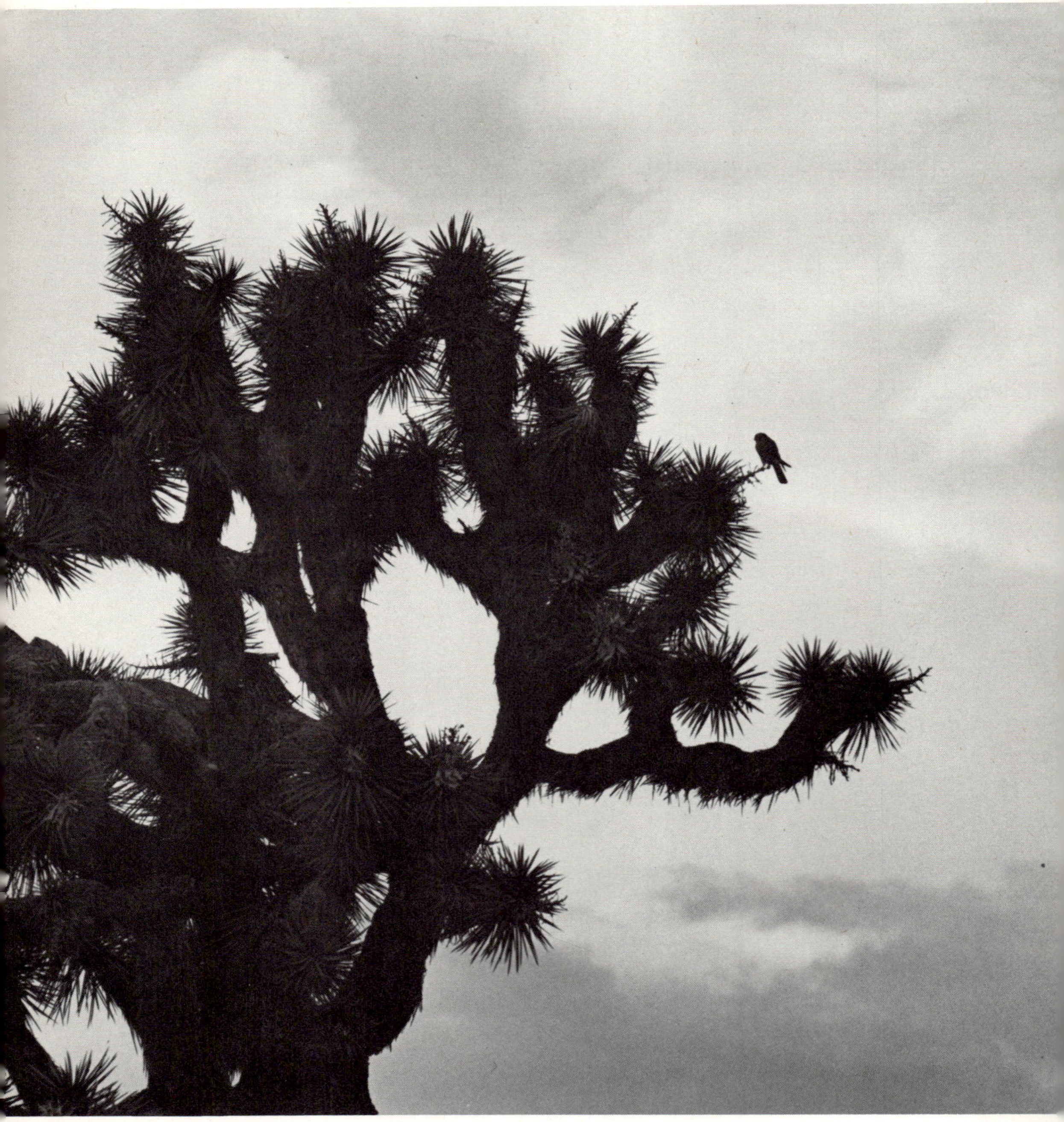

What we know so far about Joshua tree, saguaro, and sagebrush ecosystems is probably only a small part of what there is to learn. We know much more about forests, lakes, and other ecosystems than we do about deserts. There are still many mysteries of desert ecology—the relationships between living things and their environment.

These relationships often turn out to be more complicated than they were first thought to be. Take, for example, the large areas of creosote bushes that grow in the Mohave Desert. The bushes are spaced well apart from each other. In places they look as if they were planted at regular intervals, like trees in an orchard. They grow this way naturally, and people have wondered why.

Perhaps the reason is competition for water. The root system of the creosote bush is believed to capture moisture from near the soil surface, all around the plant. The open spaces between bushes were thought to represent the area of soil where root systems quickly took up any available water. This prevented new plants, including young creosote bushes, from surviving in these spaces.

Sparrow hawks perch and nest in the eerie-looking Joshua trees.

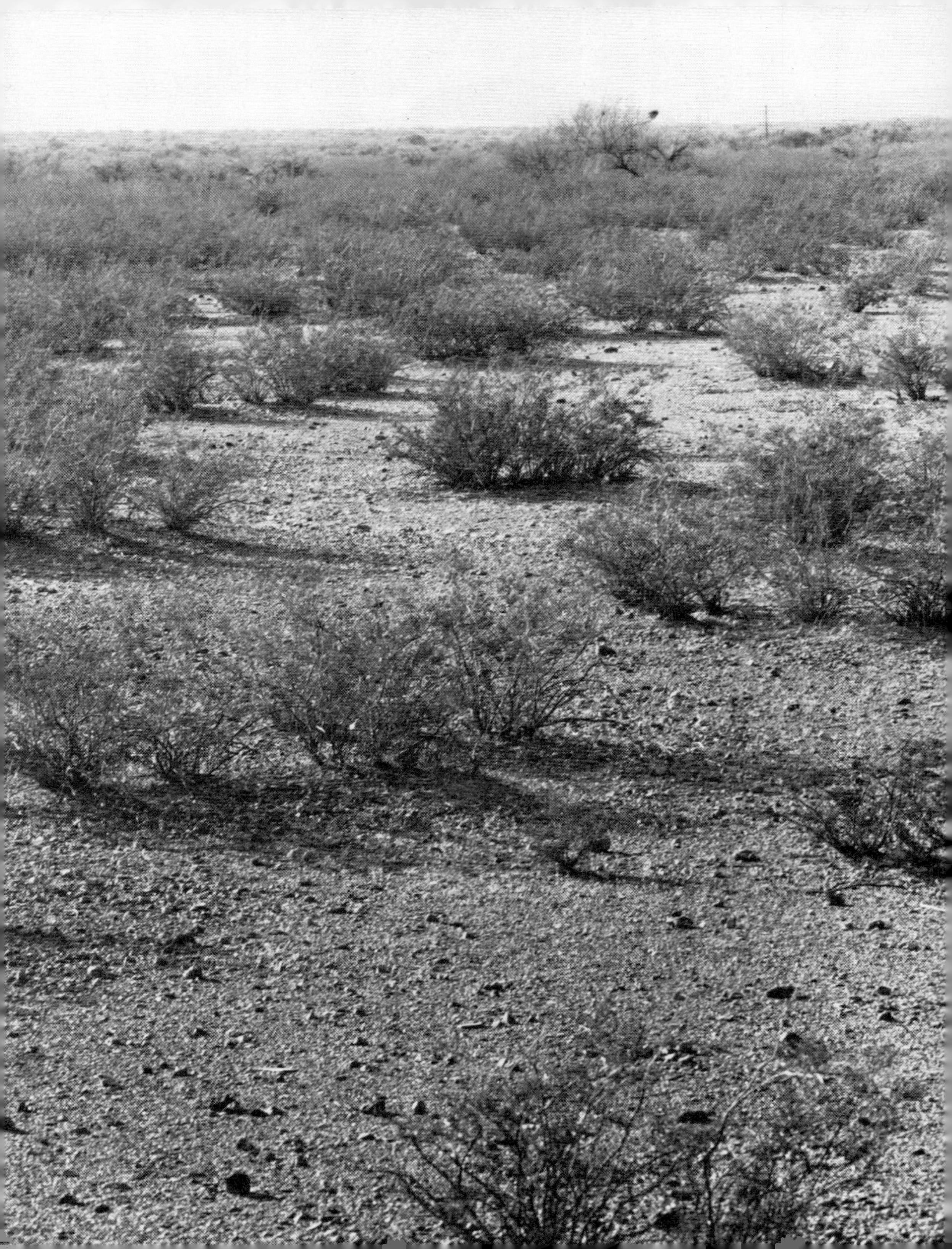

Later, another reason was suggested for the spaces between bushes. Some people suspected that chemicals given off by the creosote bushes might act like poisons in the soil. They would keep seedlings from growing near a large, well-established plant. However, biologists have made extracts of the chemicals in creosote leaves and twigs. They have watered seeds and seedlings of creosote bushes with this "poisonous" extract. The seeds sprouted normally, and the seedlings grew well.

Meanwhile, a myth about creosote bushes has been created. Many books about deserts have reported the regular spacing of these shrubs. People have been given the idea that creosote bushes grow this way everywhere in the desert.

In 1973 this notion was challenged by a botanist from the University of California. Dr. Michael Barbour gathered information and looked at creosote bushes from 42 different sites in the Mohave, Sonoran, and Chihuahuan Deserts. From these studies he concluded that regular spacing of these bushes is rather uncommon. The bushes also grow in clumps of several plants, or in a random pattern. Why the bushes have different growth patterns in different places is still a mystery. Competition for water is probably only part of the answer.

Creosote bushes cover millions of acres of the North American Desert.

Scientists usually find that nature is more complex than people first thought. We can expect this to be true of deserts. For many years people thought that cactus plants had spines for just one reason: protection against plant-eating animals.

The spines may offer some protection, but it is far from complete. Jack rabbits, ground squirrels, and several other plant-eaters are not repelled by cactus spines. Pack rats, also called desert wood rats, are often most abundant in dense growths of a cactus called jumping cholla (pronounced CHOY-ah). The cactus is named for its spiny "joints," which stick so easily to shoes, clothes, and skin that they seem to jump onto passersby. (Some desert hikers carry pliers with which to rid themselves of the pesky cholla.) Pack rats often make large nests from several bushels of cholla joints. In the Sonoran Desert about half of their diet is cactus.

Nevertheless, spines probably protect cacti from some animals. Surprisingly enough, they also provide shade. On heavily spined cacti they may cast shadows on a quarter of the plant's surface at a time. They also slow the force of the wind before it reaches the plant's surface. In both of these ways the spines may help keep cactus plants from losing precious water to the air. They may also serve other purposes, yet to be discovered.

> Cholla cacti are abundant in the Sonoran Desert. Pack rats make fortress-nests (upper right) from the spiny cholla joints. The joints stick so easily and tightly to cloth and skin that it is wise to remove them with pliers.

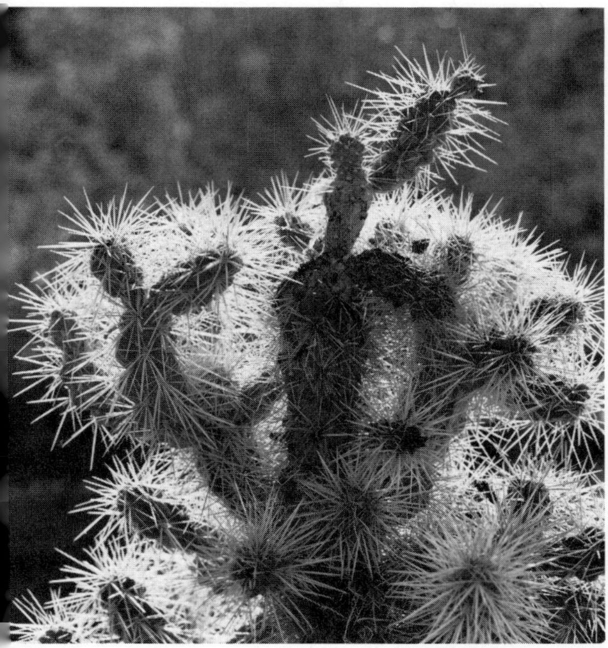

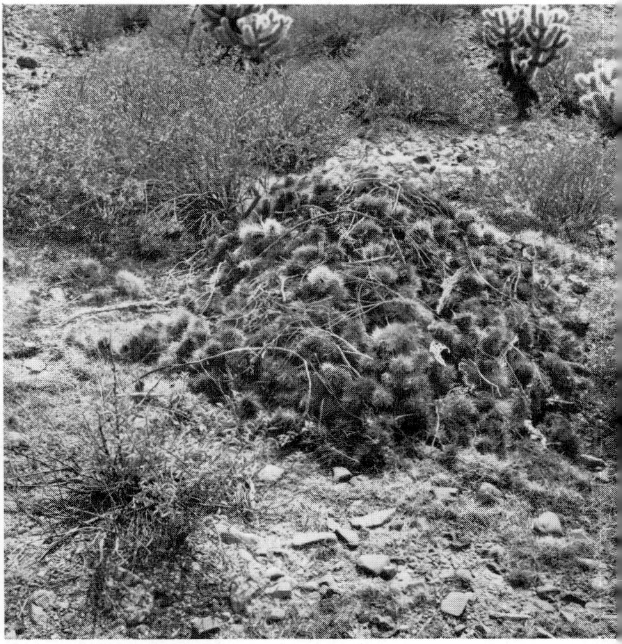

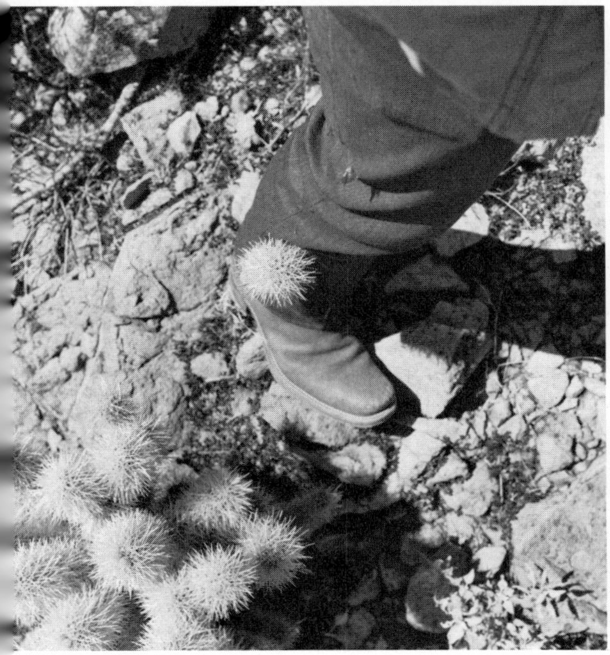

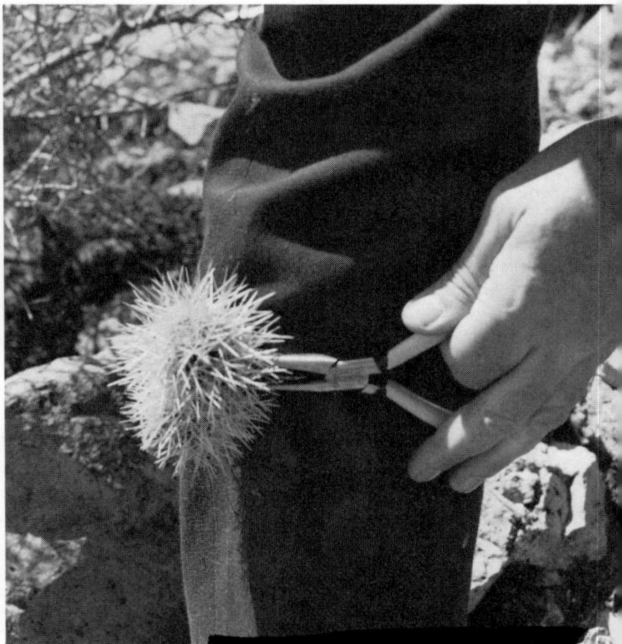

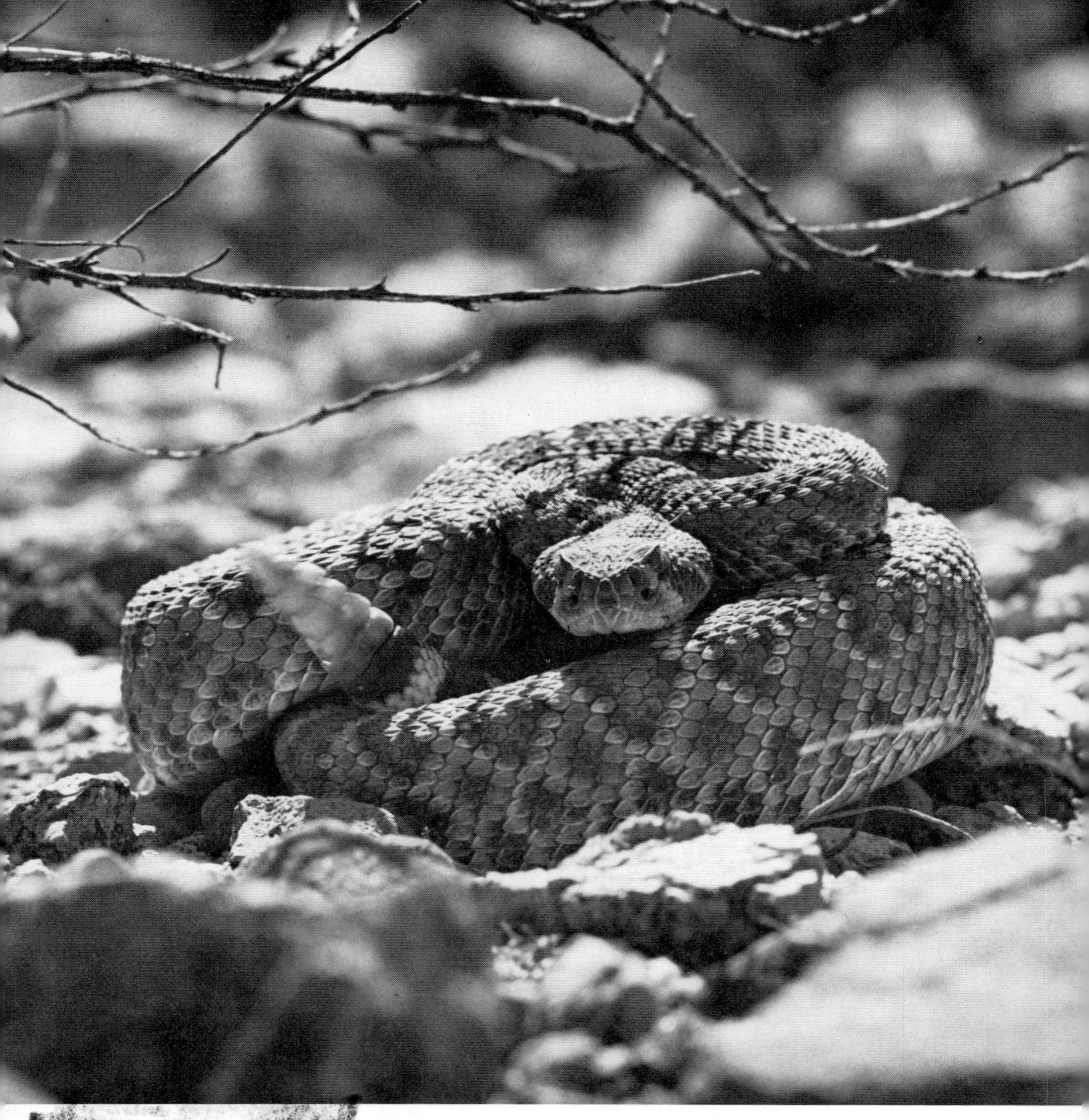

During the day, a desert can be terribly hot. People wonder how plants and animals live there. In a way, desert organisms manage in the same way people cope with the problems in a tough city neighborhood. In order to feel safe, people living there must adjust to the conditions around them. They try to lessen the chances of being robbed by not walking in certain areas at night, or perhaps at any time. Eventually it becomes routine to have several door locks, or for children to hide school lunch money in one of their shoes.

Deserts can also be tough neighborhoods. A rattlesnake may die if it is forced out into the desert sun. Kangaroo rats, exposed to air temperatures of 110° F (43° Celsius) or more, die within an hour and a half. Yet these animals, and many other organisms, thrive in the desert. It has been their home for many thousands of years. They routinely adjust to conditions that would kill other life. They have adapted to the desert environment in many intricate and fascinating ways.

Desert rattlesnakes are most active at night, and hide in shady places during the day.

Lack of water makes a desert, and lack of water is a problem that desert plants and animals have "solved" in order to survive. Many animals, including people who hike in the desert, help solve this problem by simply staying out of the sun during the hottest part of the day. But desert plants must stay in their places. The long-lived plants face several months without rain, and fierce daytime temperatures.

Plants have adapted to these conditions by using, and losing, as little water as possible. Normally, plants lose a great deal of water from little openings in their leaves. Many desert plants have waxy coatings on their leaves which help reduce this loss. They also shed their leaves during the driest times of the year. Palo verde trees, ocotillo (oh-koh-TEE-oh) bushes, and some other plants can keep making food even after their leaves have been shed. Their bark contains chlorophyll, the green substance that is vital for food-making.

During drought the palo verde tree sheds its leaves, but continues to make food within its light-green bark.

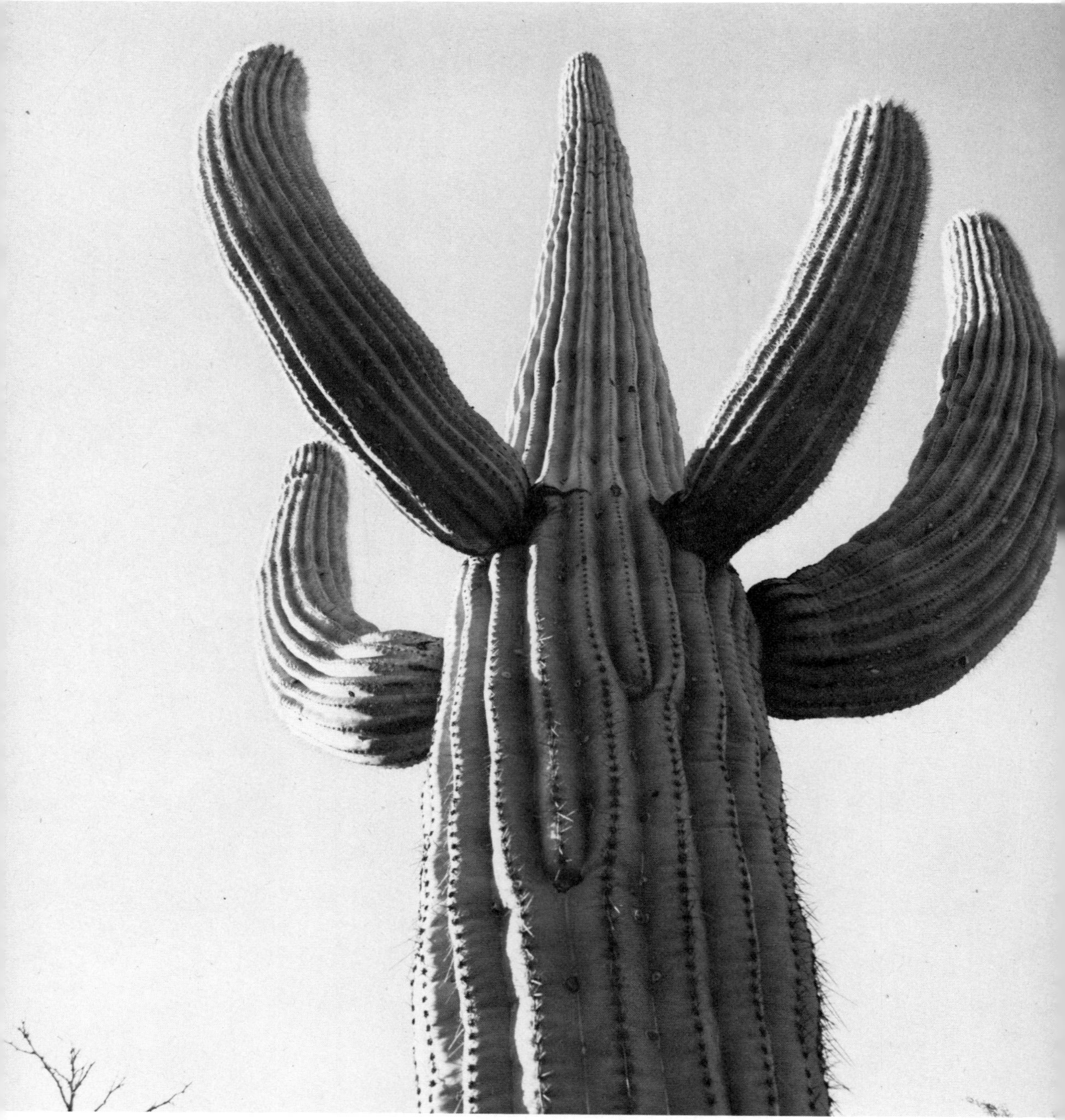

Cactus plants are symbols of the desert, and they thrive in its dryness and heat. Outside they look rough and dry, but inside they are wet and juicy. They are usually green all over, and make food without having leaves. This helps save water, and they are also able to store water. A big saguaro may store several tons after a rainstorm. Sometimes a saguaro's trunk bursts open because it cannot hold all the water its roots have soaked up. As month after month of drought passes, the stored water inside a cactus is used up, and the plant shrinks in girth.

Some desert plants avoid the challenge of drought. They exist as embryos inside tiny seeds for most of a year. Then, when enough rain falls, they sprout, grow, and blossom. Their lives are over in a few weeks. These plants are often called ephemerals, which means "living for a brief time." They leave seeds on the desert soils where, a year or more later, a new generation of plants will spring up after it rains. A light rain has no effect on the seeds. Only a heavy rainfall gives them a chemical "signal" that soon after makes the desert a garden of wildflowers.

The trunk of a saguaro (left) swells with stored water. The deep blue blossoms of Canterbury bells (right) appear after rain falls in the Mohave Desert.

Chuckwallas eat the flowers and fruit of cacti.

Desert animals are most active during the rainy season. In fact, rain triggers the emergence of insects just as it does the sprouting of seeds. Their adult lives are often as short as the ephemeral plants upon which they depend for food. Spiders, lizards, and birds feast on the abundant insects. Large lizards called chuckwallas climb into bushes and eat flowers. Many animals have their young about the time of the rainy season, when there is plentiful food for both plant-eaters and predators.

Jack rabbits emerge from their cool hideouts at dusk.

This is the desert's liveliest season. And yet you will probably see few animals at midday. They would lose water rapidly if they were active then. Instead, they rest in shady places or underground. Jack rabbits crouch or sprawl beneath bushes. They lose some body heat by breathing rapidly, or panting. They may also lose some from their large ears.

Some kinds of snakes and lizards crawl up into bushes, away from the soil surface. The temperature at ground level may reach 150° F (66° Celsius) or higher. The air is many degrees cooler just a foot above the surface. The temperature a foot below the surface—where many rodents burrow—may be a comfortable 85° F (29° Celsius).

Some animals are better able to take the heat than others. The antelope ground squirrel often remains active after other animals have retreated to shelter. Finding a cool patch of soil, the squirrel presses the underside of its body there for a few moments. The squirrel is cooled. In this way the squirrel can lower its body temperature several degrees in a few minutes. Then it is ready for more activity in the sun.

The color of desert animals also helps them avoid overheating. Nearly all of the desert landscape is light-colored, and so are nearly all desert animals. Light colors reflect sunlight, while dark colors absorb it. For most of the year, desert animals must avoid getting too warm, so their light colors are an advantage. (For the same reason, desert visitors are advised to wear light-colored clothes.)

Light colors probably give animals some protection from predators. A light-colored mouse, insect, or lizard is difficult to see against a light background. Most dark-colored animals are found only on areas of black lava and volcanic cinders. A lava area in New Mexico is the home of some dark-colored lizard species. The same kinds of lizards live a few miles away at the White Sands National Monument, where they are light-colored. One of the whitest of all desert animals—the little Apache pocket mouse—lives among the white dunes and nowhere else in the world.

The color of desert grasshoppers usually matches the ground where they live.

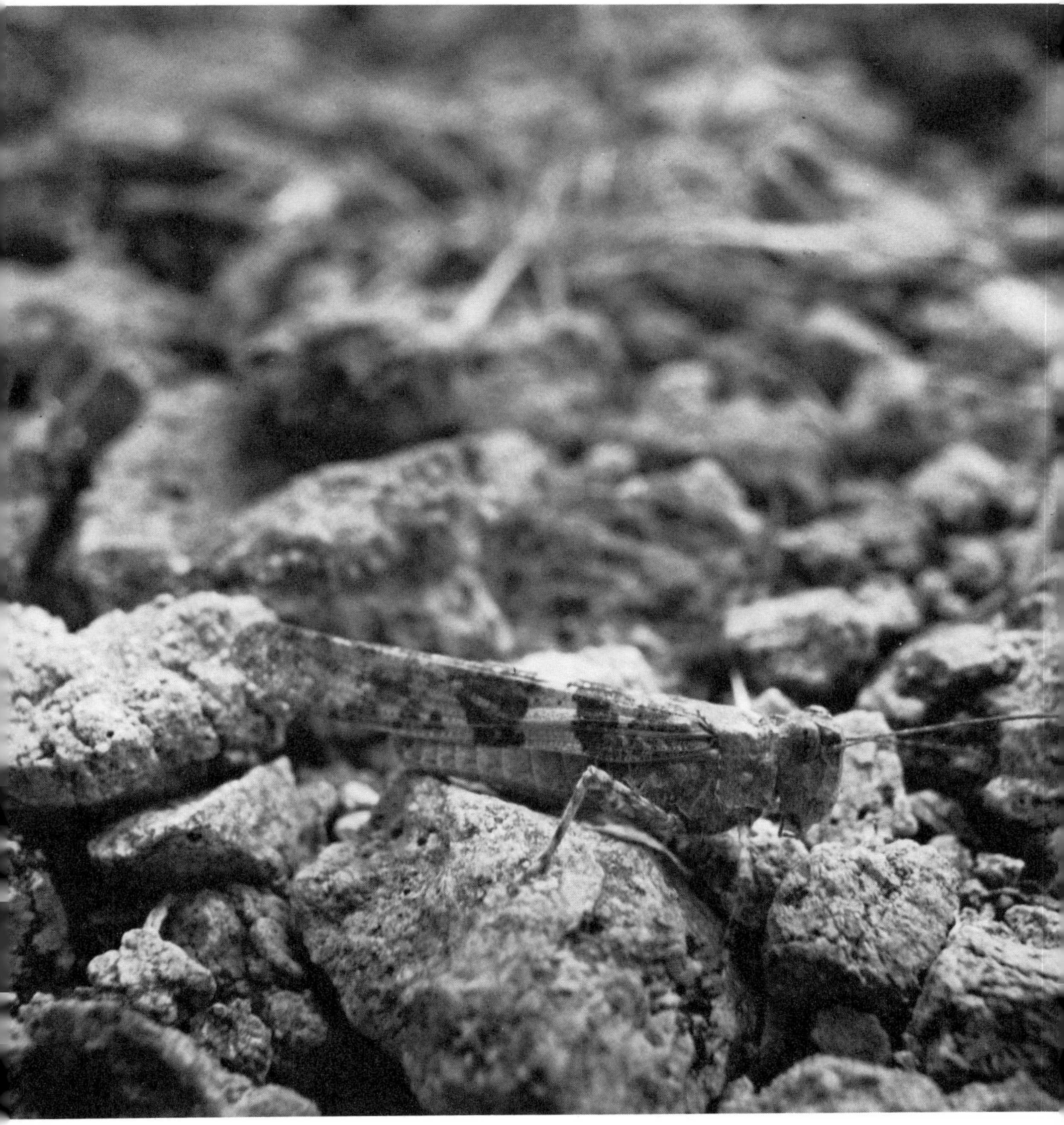

Every living thing needs water in order to keep alive. Desert animals are no exception. Some of them lap dew from plants in the morning. The larger mammals and birds may travel some distance to a water hole. For many, however, there is no year-round supply of drinking water. Often their food is a major source of moisture. Every insect, mouse, and leaf is at least half water.

Several species of kangaroo rats never drink. Their needs are met by water that is produced as seeds are digested in their bodies. All animals, including humans, produce some water this way. But only kangaroo rats can live on this metabolic water, as it is called. They need very little liquid because they lose very little when wastes leave their bodies. Also, like most other rodents, they rest in burrows during the heat of the day, and emerge to find food at night.

The desert night is busy with hopping, scurrying rodents. Almost never seen by visitors, these small mammals are a vital part of desert ecosystems. They convert plant food into animal flesh. Then many of them become food for snakes, foxes, owls, and other creatures of the night. One group of rodents are predators themselves. They are called either grasshopper mice or scorpion mice, and are named after some of their common foods.

Bobcats (top) rest out of the sun during the day, while kangaroo rats (bottom) find a cool microclimate underground.

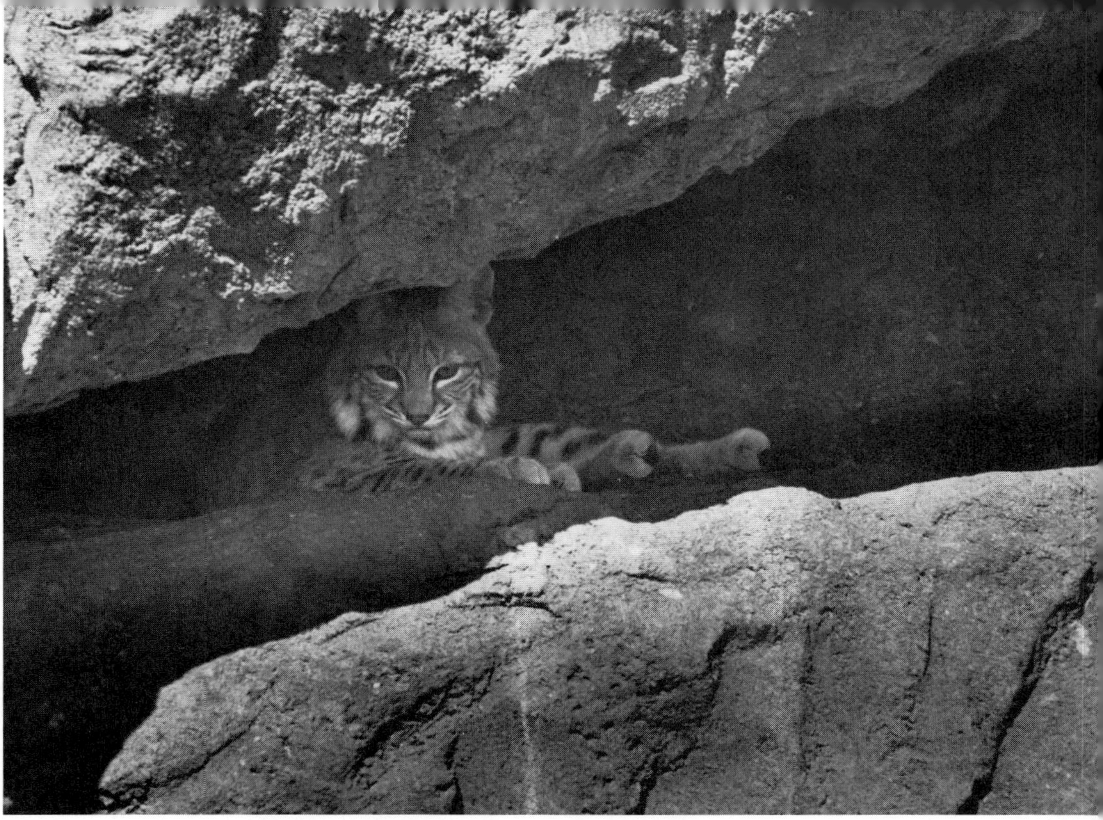

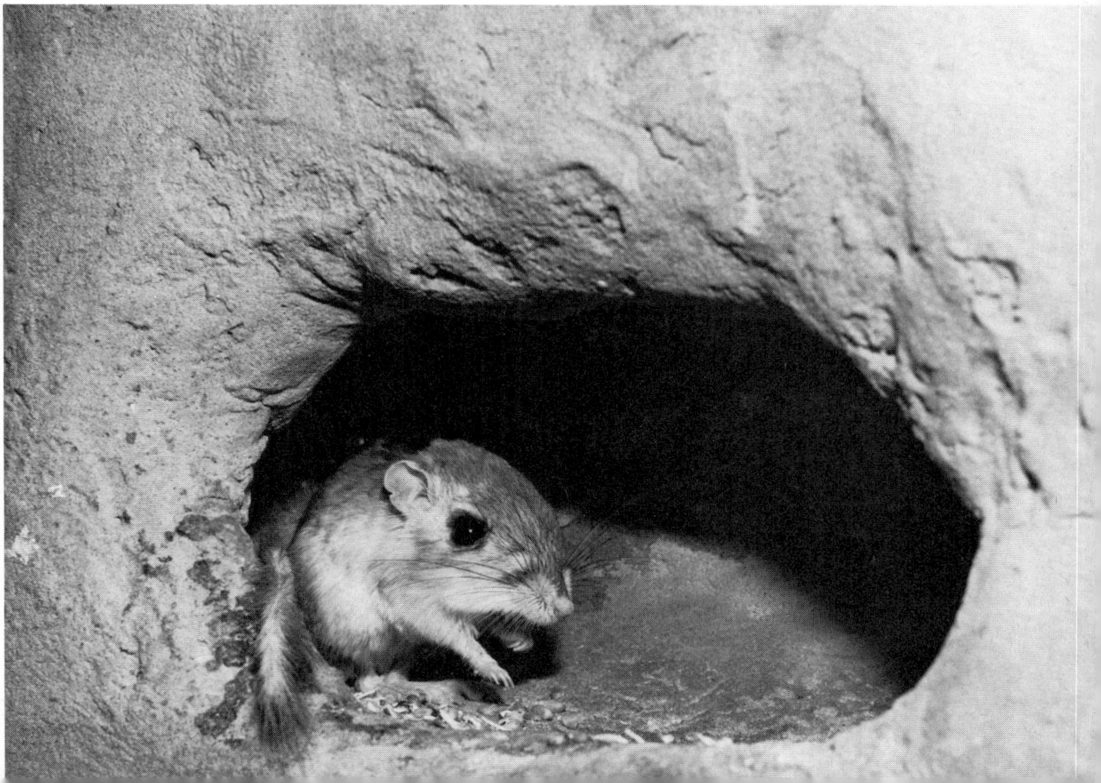

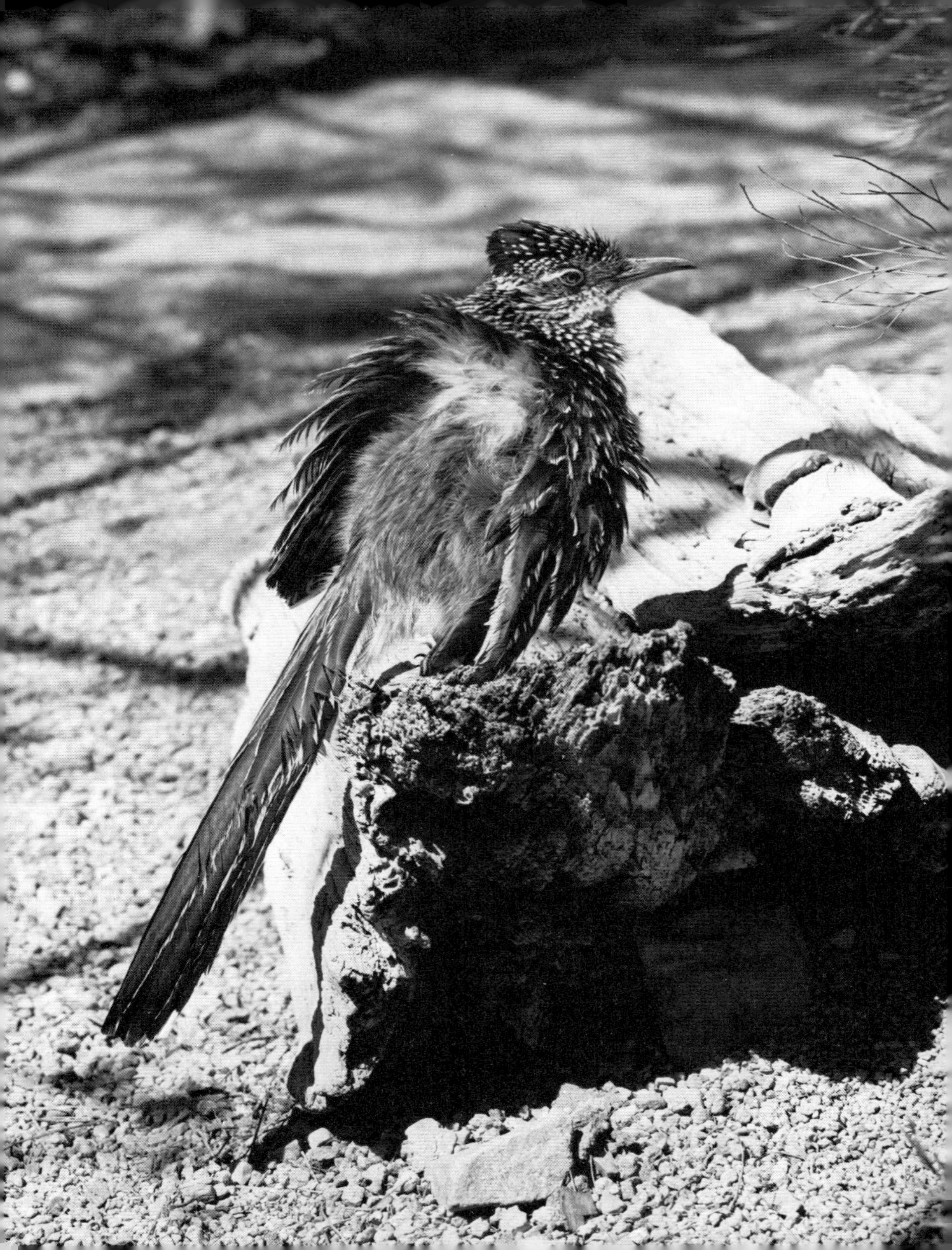

At night the desert air cools sharply. By dawn some animals face the problem of getting warm. Horned lizards flatten their bodies as much as possible, and turn at right angles to the sun's rays. In this way they expose much of their body surface to the sunlight. The big, comical birds called roadrunners may also need to warm up. They turn their backs toward the sun and raise their body feathers, exposing black skin which absorbs heat from the sun. After soaking up enough warmth, they lower their back feathers and dash off in pursuit of lizards and small snakes.

The roadrunner is the official state bird of New Mexico. This fascinating bird, rather than a barren sand dune, is also becoming a symbol of the North American Desert—the ecosystem which people are finally getting to know. Yes, the desert can be very hot. Some poisonous animals live there. It looks very different from the forests and lawns most people know best. But any place that is home for roadrunners, antelope ground squirrels, saguaros, and fringe-toed lizards must be quite special and worthwhile. For these organisms and thousands of others, the desert is a gentle and bountiful place.

With its back feathers raised, a roadrunner warms its body in the morning sun.

Not everyone who visits the desert or lives there feels this way. It seems that most people still look upon the desert as a place to be used and changed. It has been grazed—and overgrazed—by livestock. If the desert in some areas seems barren, the barrenness is often manmade. Many thousands of acres have been strip-mined for coal. No one knows whether this land will ever recover and once again be a complete desert ecosystem.

Many thousands of motorcycle riders appreciate the desert, but not for its beauty and wildlife. In 1973, about seven hundred motorcyclists raced illegally through a part of the Mohave Desert that had been set aside as a reserve for the rare desert tortoise. According to desert biologists, the plant life where the motorcycles passed will not recover in this century.

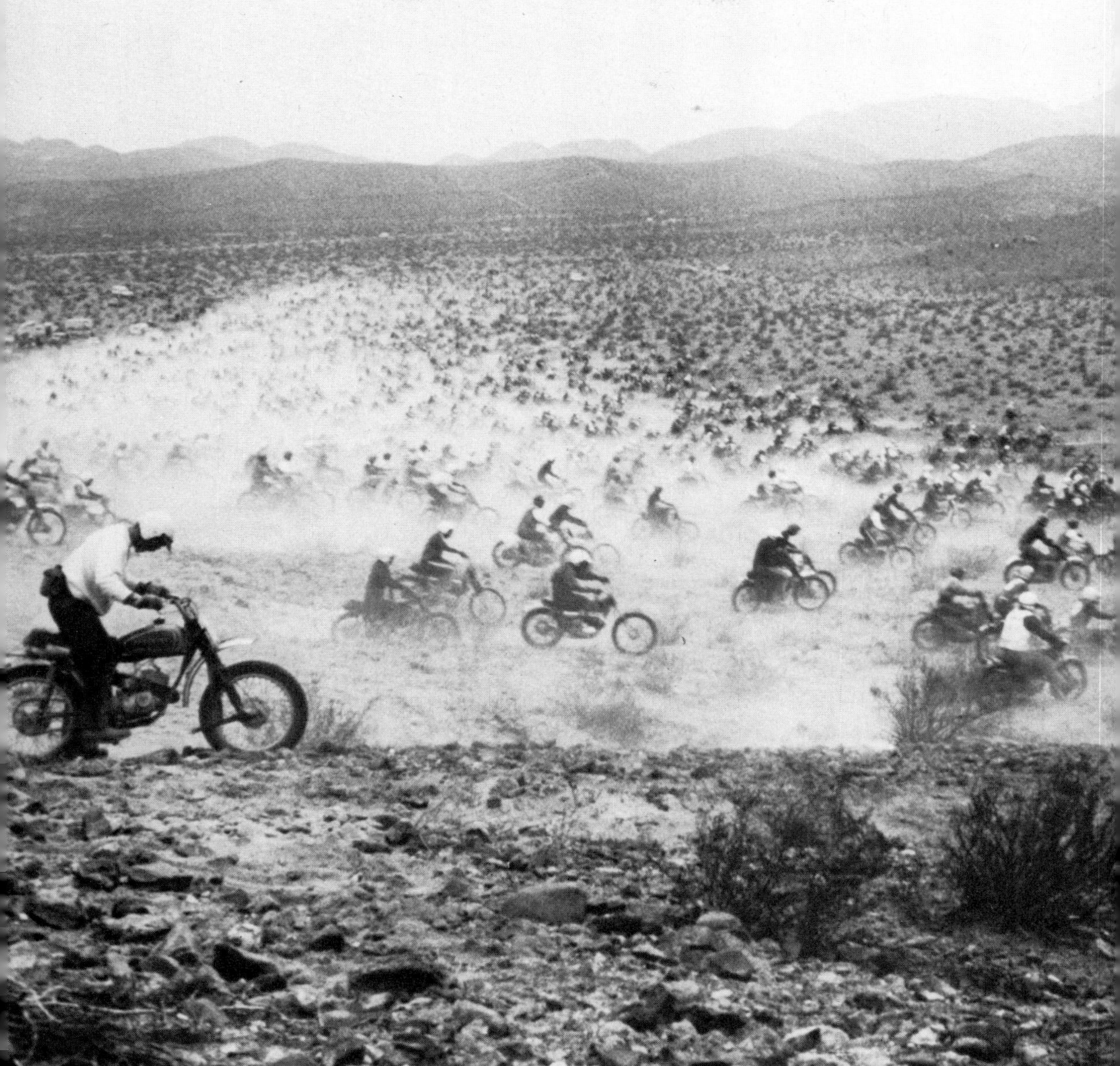

CLOSED
AREA
THE AREA BEHIND SIGN
CLOSED
TO ALL VEHICLE USE
TO PROTECT NATURAL VALUES

The motorcycle race is just one example of a very big problem. In California alone there are an estimated 200,000 dune buggies and a million motorcycles designed for off-road use. Many of the owners of these vehicles view the Mohave Desert as their playground. But most of this desert is public land. It belongs to all the citizens of the United States. The Bureau of Land Management oversees this land. It has the tough job of protecting the desert and also allowing people to use it for recreation, grazing livestock, and other purposes.

Large areas of the North American Desert are owned by people who are free to do whatever they want with it. Millions of acres are now covered with homes, factories, shopping centers, roads, and farmland. The population of the southwestern United States is growing rapidly. The people there need food, electricity, and water.

Signs alone are obviously not enough
to protect public lands from damage by
off-road vehicles.

The Colorado River and other rivers that flow through the Southwest provide only part of the water that is used. In Arizona most of the water is pumped from underground supplies, but these supplies are limited. No one knows what the limits are. Some ground water in the Southwest is thousands of years old. Once it is gone, it may not be replenished for centuries. Each year water levels drop lower, and wells must be drilled deeper and deeper.

There are schemes for bringing water from other regions. There are hopes of changing huge amounts of salty ocean water into fresh water. And there are people who suspect that parts of the Southwest will run short of water long before such things happen. Already some farmland in Arizona has been abandoned for lack of water.

Scarcity of water in the Southwest may bring an end to showy fountains and other waste.

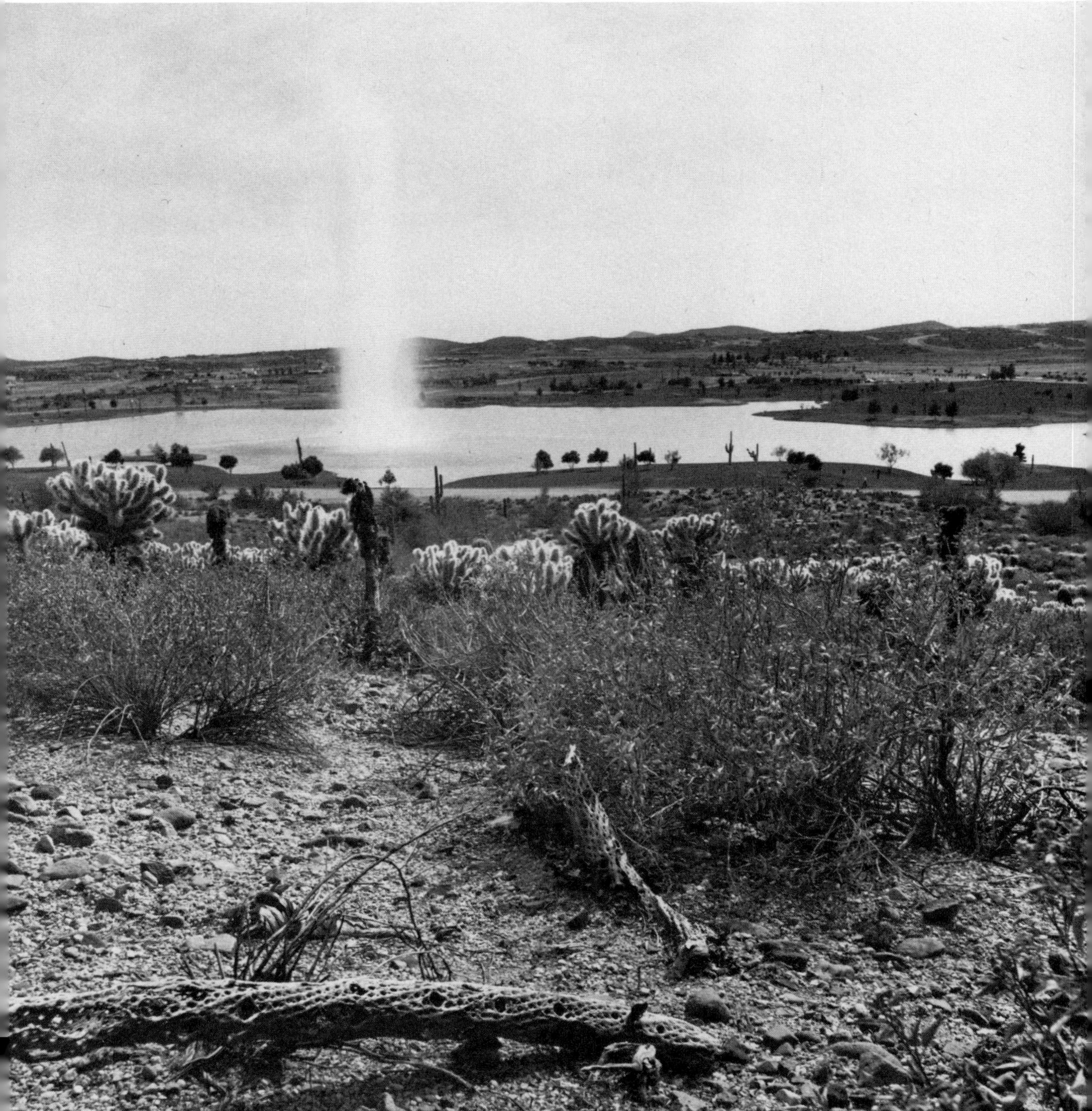

The effects of people on deserts reach far beyond cities, farms, and roads. In Death Valley, desert holly plants are dying. A plant ecologist believes the cause is polluted air from Los Angeles, almost two hundred miles away.

Nevertheless, there are still vast areas of wild desert. Parts of the North American Desert may be among the least damaged ecosystems on earth.

There are people who want to "improve" this land, to change it, to "make the desert bloom." There are other people who have learned to appreciate the desert just as it is. They feel that the desert has always bloomed, in its own special way. They are working to save the gentle desert.

Glossary

ARROYO—a desert stream channel which is dry for most of each year, carrying water only after a heavy rain. Also called a dry wash.

CHLOROPHYLL—the green colored substance in plant cells that is needed for the food-making process called photosynthesis, in which sun energy is converted to food energy.

ECOLOGY—the study of relationships between living things and their environment.

ECOSYSTEM—a place in nature with all of its living and nonliving parts, including soils and climate.

EMBRYO—a plant or animal in its earliest stages of development, before birth, hatching, or sprouting (of a seed).

EPHEMERALS—plants or animals that live only a short time in their fully developed, mature form. This term is usually applied to some kinds of insects and to many kinds of desert wildflowers.

GROUND WATER—water from rain and melted snow that soaks underground into sand, gravel, and certain kinds of rocks. The water pumped to the surface from wells is ground water.

HARES—mammals that are closely related to rabbits, but which have longer ears and longer hind legs. Also, young hares are born well-furred, and with their eyes open. They are soon able to fend for themselves. Young rabbits are born naked, with eyes closed, and cannot survive without parental care.

LARVAE—the young of some groups of animals, especially insects. Larvae are usually an active stage of insect development; a caterpillar is the larva of a butterfly or moth.

LICHENS—plants that are "partnerships" between fungi and algae. The fungi provide support and trap water that is used by the algae. Food made by the algae is used by the fungi. Lichens often grow on the surfaces of rocks and tree bark.

METABOLIC WATER—water that is produced during food digestion within animals. As sugar is broken down chemically, hydrogen and oxygen are produced. They combine to form metabolic water. Ordinarily this water provides only a small part of an animal's needs, but it helps some desert rodents live without drinking.

MICROCLIMATE—"little climate." The climate (including temperature, amount of water vapor in the air, winds) in a small area. Great differences in climate occur within a short distance: for example, at the soil surface in the desert and a few feet above the surface.

PLAYA—a temporary desert lake. Like arroyos, playas usually contain water for a short period following a heavy rain. The water evaporates into the air; then the playa is dry until the next rainfall, a year or more later.

POLLINATION—the process by which pollen from one flower is transferred to another, so that the egg cells are fertilized by sperm cells in the pollen and new embryo plants (seeds) are produced.

POLLUTION—people-produced wastes, such as heat, noise, sewage, and poisons, that lower the quality of the environment.

PREDATORS—animals that kill other animals for food.

RODENTS—gnawing mammals which have four prominent yellow or orange teeth which grow throughout their lives. Rodents include mice, rats, lemmings, beavers, and squirrels, but not rabbits and hares.

Further Reading

Books and periodicals marked with an asterisk (*) are fairly simple; the others are more difficult.

ABBEY, EDWARD. *Desert Solitaire: A Season in the Wilderness.* New York: McGraw-Hill Book Company, 1968. (Also available in paperback.)

———. *Cactus Country.* New York: Time-Life Books, 1973.

BARBOUR, MICHAEL. "Desert Dogma Reexamined: Root/Shoot Productivity and Plant Spacing." *American Midland Naturalist,* January 1973, pp. 41–57.

* BRENNER, BARBARA. *Lizard Tails and Cactus Spines.* New York: Harper & Row, 1975.

BROWN, JAMES, and LIEBERMAN, GERALD. "Woodrats and Cholla: Dependence of a Small Mammal Population on the Density of Cacti." *Ecology,* vol. 53, no. 2 (1972), pp. 310–313.

COSTELLO, DAVID. *The Desert World.* New York: T. Y. Crowell Company, 1972. (This book concludes with a helpful section on "how to see the desert.")

FINDLEY, ROWE. *Great American Deserts.* Washington, D.C.: National Geographic Society, 1972.

JAEGER, EDMUND C. *Desert Wildflowers.* Stanford, California: Stanford University Press. Revised edition, 1941. (Paperback)

———. *Desert Wildlife.* Stanford, California: Stanford University Press, 1961.

KIRK, RUTH. *Desert: The American Southwest.* Boston: Houghton Mifflin Company, 1973.

LARSON, PEGGY. *Deserts of America.* Englewood Cliffs, New Jersey: Prentice-Hall, Inc., 1970.

———. *The Sierra Club Naturalist's Guide to the Deserts of the Southwest.* San Francisco: Sierra Club Books, 1977. (Also available in paperback.)

* LAUBER, PATRICIA. *Life on a Giant Cactus.* Champaign, Illinois: Garrard Publishing Co., 1974.

* LAVINE, SIGMUND. *Wonders of the Cactus World.* New York: Dodd, Mead & Company, 1974.

LEOPOLD, STARKER. *The Desert* (Life Nature Library). New York: Time-Life Books, 1961.

* PITT, VALERIE, and COOK, DAVID. *A Closer Look at Deserts.* New York: Franklin-Watts, 1975.

STEBBINS, ROBERT, and COHEN, NATHAN. "Off-Road Menace: A Survey of Damage in California." *Sierra Club Bulletin,* July–August 1976, pp. 33–37.

* SUTTON, ANN, and SUTTON, MYRON. *The Life of the Desert.* New York: McGraw-Hill Book Company, 1966.

* VENNING, FRANK. *Cacti.* New York: Golden Press, 1974.

Index

An asterisk (*) indicates a photograph or drawing

air pollution, 52
antelope ground squirrels, 40, 45
ants, 22
Apache pocket mouse, 40
Arizona, water use in, 50, *51
arroyo (dry wash), 8, *9, 17

badgers, 24
Barbour, Dr. Michael, 29
bats, 21, 22
birds, 3, 11, 21, 22, 24, 38, 42; see also hawks, owls, quail, roadrunners, woodpeckers
bobcats, *43
Bureau of Land Management, 49

cacti, 8, *17, 30, *31, *36, 37; see also cholla, saguaro
chlorophyll, 34
Chihuahuan Desert, *7, 8, 29
cholla cactus, *31
chuckwalla lizards, *38
cockroaches, 14
Colorado River, 50
colors of desert animals, 40
coyotes, *10, 11, 22
creosote bushes, 27, *28, 29

Death Valley (California), 6, 52
deer, 11

deserts: causes of, 4, *5; characteristics of, 1, 4, 6, 8; climate of, 1, 4, 6, 18, 33, 39, 45; North American, 6, *7, 12, 45, 52; polar, 6
doves, 11, 22
dunes, 1, 12–14, 17, 40, 45

ecology, 27
ecosystems, 1, 12, 17, 18, 27, 42, 45, 46, 52
ephemeral plants, *37, 38

food chain, 22
foxes, 42
fringe-toed lizards, 1, 14, *15, 45

gophers, 24
grasshoppers, *41
grazing by livestock, 46, 49
Great Basin Desert, 6, *7, 8, 24

hawks, 18, *20, *26
holly, desert, 52
horned lizards, 45

insects, 3, 11, 14, 21, 24, 38, 40, 42
ironwood, 8

jack rabbits, 3, 11, 22, 30, 38, *39

57

Joshua tree, 24, *26, 27

kangaroo rats, 24, 33, 42, *43

lichens, 17
lizards, 3, 14, 21, 38, 39, 40, 45;
   see also individual species

metabolic water, 42
mesquite, 8
mice, 21, 22, 24, 40, 42
microclimate, 18, 21, 39
Mohave Desert, *7, 8, 24, 27, 29,
   46, 49
mosquitoes, 21
moths, 12, 21
motorcycles, 46, 49

ocotillo bushes, 34
owls, 1, 18, *21, 42

pack rats, 21, 30; nests of, 30,
   *31
palo verde tree, 8, *19, 34, *35
pickleweed, 17
playa, *16, 17
predators, 14, 21, 38, 42

quail, *11

rattlesnakes, *32, 33
roadrunners, *44, 45
rodents, 11, 14, 22, 39, 42; see also
   individual species

sagebrush, 8, 24, 27
saguaro cactus, 8, 18, *19, *20,
   21–22, 24, 27, *36, 37, 45
salt bush, 17
scorpions, 1, 21, 42
seeds: of arroyo plants, 11; of
   creosote bushes, 29; of ephemeral
   plants, 37, 38; of saguaros, 22;
   of yuccas, 12
smoke tree, 8
snakes, 14, 24, 33, 39, 42, 45
Sonoran Desert, *7, 8, 18, 24, 29,
   30
spiders, 21, 24, 38
strip-mining of coal, 46

tortoises, desert, *46

vultures, turkey, 18, 22, *23

wash, see arroyo
water: competition for, 27, 29;
   saving of, 30, 34, 37, 39, 42;
   sources of, 11, 17, 21, 50
White Sands National Monument
   (New Mexico), 12, 40
woodpeckers, 21, 22
wood rats, see pack rats

yucca, 12, 24
yucca night lizards, 24, *25